FRANCIS BUCAILLE

THE INTELLIGENT SOIL

Restoring Fertility, Plant Health and Resilience

Translated by Amanda Liles-Abiven

ORIGINAL FRENCH EDITION :
« *Revitaliser les Sols* »
© Dunod, 2020
11 rue Paul Bert, 92240 Malakoff
www.dunod.com
ISBN 978-2-10-080918-9
Legal deposit: October 2020 - Continued printing: January 2022
Revised and updated French edition 2025

First edition 2025

Cover photo ©
Top Front page from left to right : F.Bucaille, soil test for nitrate leaching ; iStock wheat field at sunset, F.Bucaille lavender field in Provence (France), F.Bucaille Vineyard in Corsica
Bottom Front page : F.Bucaille. The author analyzing a soil profile in California under citrus orchard
Back page : F.Bucaille TEDx Talk

Code ISBN : 979-10-979413-0-7

ACKNOWLEDGEMENTS

Thank you to all of the authors, thinkers and agronomists, living or deceased, who have contributed to the construction of the field of agronomic science and my own vision, with special mention to Neal Kinsey who accompanied me with generosity and kindness.

Gratitude is also extended to all my fellow farmers who, through their observations, relevance, and audacity, have allowed me to progress. Much of what is recounted here is merely a compilation of the generous sharing of their discoveries. Thirty of them, my 'Reference Farms' in France and Belgium, cereal growers, breeders, fruit producers, vegetable growers, and winegrowers, have been exceptional contributors. They are to be thanked for their past, present, and - I count on them - future contributions. They will recognize themselves in this collective address.

Finally, many thanks to the translator: Amanda Liles-Abiven

TABLE OF CONTENTS

PREFACE

Caring for the soil we owe to our children

I'm a major part of the earth's biodiversity, I govern the cycles of matter, I'm a source of food and I play with the greenhouse effect. Who am I?

The soil. More than just the support for our steps or for plants: it's an intense place of life. A gram of soil hosts over a billion bacteria, of numerous different species; it also includes 1 to 100 thousand species of fungi...the soil bacteria in France alone encompass at least 115,000 species, compared to 570 birds, 6,500 plants, or 189 mammals! With 50% of living biomass, 23% of known living species, and 75% of terrestrial organic matter, the soil IS the terrestrial ecosystem: what we see on the surface is only a diverticulum!

Life creates the soil: it breaks down organic matter to recycle its elements; it attacks rocks to release fertility; it exploits the atmosphere, with gaseous nitrogen transformed by bacteria into organic nitrogen. Life stirs the soil, with animal movements and the uptake of elements brought up from deep by plants. Moreover, 90% of plants cannot survive without the so-called "mycorrhizal" fungi, which, on one hand, search for resources in the soil and, on the other hand, colonize roots to provide minerals collected from the soil. Long seen as an interface between the geosphere and the atmosphere, the soil is not only an ecosystem but above all a living process.

Soils shape the world. Their fertility carried away by water fertilizes the oceans, explaining why waters near continents are the most productive (even fishing comes from soils!). Soil organisms emit greenhouse gasses: CO2 from the respiration of aerated soils, methane, and nitrous oxide from anaerobic soils warm the planet. Conversely, reasonably aerated soils degrade organic matter slowly: a simple solution against the greenhouse effect is to bury our organic waste in the soil! Increasing the organic matter content in soils by 0.4% per year would store the equivalent of our annual CO2 emissions!

Sadly, humanity has not understood this. Development and urbanization cover the soils of a department every 7 to 10 years in France. Salinization threatens 30% of agricultural soils because irrigation brings salts that accumulate. Plowing brings fertility to the surface, aerates the soil, and weeds our fields, but its indiscriminate practice (repetitive and deep) harms in the long term. Conversely, when practiced in autumn and without plant cover, it boosts winter erosion of bare soils, with exploded structure, where oxidized organic matter no longer provides cohesion: our plowed soils erode on average ten times more than before the plow.

Soils are damaged at their own pace, a slow but certain pace. It takes 100 to a thousand years to create a soil; therefore, we sadly realize that they are a heritage that cannot be replaced. We inherit soils from our ancestors and we owe them to our children; we have a moral duty to use only the interests of them: a usufruct.

Catastrophic and moralistic vision? Not at all, because the book that follows offers solutions and actions. The promise fulfilled by this work is to connect the recognition of the soil's life logic with practical field work. In our developments, we are obliged to touch and modify the soils, as it is related to the nutrition of humanity. Therefore, we need practical and realistic solutions to do so while respecting the unique dynamics and sustainability of soils, to make them more fertile and resilient

Together, farmers, agronomists, and citizens, whether consumers or decision-makers, let's take hold of the soil and its ecology as levers to lift the future. It is a citizen's action to do so, an action that can only be collective. Let us remember to always question ourselves, for no practice is without side effects, no imperative holds forever without the test of facts: let's pilot together, attentive at every moment, the future of our soils. And let's start by reading the following lines, where Francis shows us the way, and the first steps that the current state of knowledge recommends...

Marc-André Selosse

Professor at the National Museum of Natural History

Professor at the Universities of Gdansk (Poland) and Kunming (China)

Member of the French Academy of Agriculture

FOREWORD

All solutions devised by living organisms have stood the test of time, thereby legitimizing them. Life likely emerged on Earth in the form of a bacterium, and these original building blocks of life are still present in contemporary biology. Since then, life has become more complex and has learned to breathe, swim, fly, communicate, manufacture materials with incredible and unmatched qualities. As a farmer, this realization fosters humility. It requires us to transition from being merely "soil miners" to "stewards of nature". We have to stop extraction, abusive use, and depletion from nature and the planet's reserves without balanced reciprocity. Whereas we are indebted to it, we are its "stewards." Stewardship implies the notion of exchanges and obligations for humans, for farmers toward nature, obligations of maintenance, of "husbandry of the fields" in this "theater of agriculture" of the 21st century (O. de Serres, *Le théâtre d'agriculture et mesnage des champs* [*The Theater of Agriculture and Field Management*], 1600). This means that the Earth is only leased to us, lent to us in exchange for respect and good maintenance. We do not "make plants grow." In fact, they agree to grow under conditions chosen by humans in places chosen by humans. Culturally, for centuries, we have been accustomed to imposing, dominating nature, even so presumptuously, to improve it. The mystery of life has been reduced for some to a double helix in color marked with ACGT codes (the nitrogenous bases of DNA) now decrypted. For a few decades, humans have started to "tinker" with their genome, sometimes even deluding themselves into believing they are participating in the creation of life. This sense of omnipotence must now give way to an ecological scientific culture that no longer places humans at the top of the living pyramid but, more realistically, as a dependent link. Then we can stop focusing on the spectacular. Then we can pay attention to seemingly insignificant things. Then we can focus on the life of the soil.

For millions of years, humans had a limited impact on the Earth, and respect for nature prevailed. It was common in many civilizations to ask for forgiveness from the deified nature before sacrificing an animal or cutting down a tree. Nature was seen as a deity, a whole, Gaia, and it was unthinkable to cut the hair of this goddess (the forest), or to penetrate her entrails (mining). The Industrial Revolution and Science completely desacralized this quasi-mystical relationship. Nature was reduced to an assembly of atoms, chemical reactions, enzymatic processes, and stripped of its sacred essence. The path was open for unbridled exploitation. Thus, man, the farmer, seeks to assert his intelligence and supremacy over all other forms of life by inventing his own solutions; solutions that often, however, life has already imagined in a more elegant, economical, efficient, and sustainable way. Man will begin to glimpse the right solutions when he erases from his imagination his own representation, so often reiterated, even in the scientific works of the 19th and 20th centuries, reigning at the top of the living pyramid. The world is circular, not linear. Everything man possesses, consumes, eats, builds, even in the heart of the city, is provided to him through his ecosystem. Oil, coal, ore transformed into steel, quartz melted into glass are nothing but concentrated solar energy transformed into chemical energy through photosynthesis. Without this marvel called chlorophyll, nothing would exist. His survival depends entirely on what he thinks he has the right to mistreat. To stop, as Rabelais said, "farting higher than he has the ass for" will allow him to observe, then to understand, and finally to respect ecosystems for his greatest benefit.

What can we learn from the long history of life? For millennia, the soil has self-maintained, self-recycled in balance with flora, fauna, microbiota, and climate. Steppes, savannas, prairies, forests and mangroves ... These biocenoses[1] have developed harmoniously and continuously across the earth. Their annual biomass production is often impressive (up to 30 tons of dry matter per hectare per year). Moreover, the complexity of these ecosystems does not imply fragility, but rather resilience, stability, and robust health. Disease and parasitism are confined to very narrow ecological niches. One cannot help but admire this symphony played flawlessly by billions of microorganisms, millions of insects, hundreds of different plants, where the rule is communication, collaboration, and adaptive evolution: the intelligence of life in full action!

If primary forests still seem to operate in a permanent cycle according to this model, what has happened to our agricultural or forest soils, which have been constantly impoverished and losing their native fauna, especially over the past seventy years? Regardless of latitude or production systems, human action has the same effects and common causes everywhere. Everywhere, the fertility and health of plants are poorly understood. The desire to write this book thus first arose from this observation, from the need to explain why certain paths have become dead ends and to propose the foundations of a new bio-inspired model,

---

[1]Biocenosis: the collection of living organisms (animals, plants, fungi, bacteria, etc.) coexisting within a given environment, the biotope. The ecosystem corresponds to the combination of the biocenosis and its biotope.

one that holistically[2] includes the natural intelligence of ecosystems. Our mindset has led us to invent so far, while there is still much to discover in what nature has already tested and solved. We must better understand the mechanisms of life, of the soil, of the trophic chain, not to better control and improve, but to better respect its laws, allowing us to integrate our presence and activity as best as possible into this workshop. **This science called ecology and this methodology called biomimicry consist of humbly observing the genius of life**. This is what we invite you to do. It is the necessary condition if we want to succeed in the shift of the second green revolution but also if we want to claim to be able to address the challenges of agricultural economy, climate, carbon storage in soils, maintenance of biodiversity, and more broadly, human health.

Why should cultivated plants inevitably become sick when natural ecosystems never do? Why are pastures dried out in summer while roadside edges remain green? Why are lands so fertile after fallow cultivation or pasture? And why should this fertility necessarily diminish over the years? Why, as farmers, should we necessarily do without the services that nature freely provides within ecosystems? The perceived inevitability, the "normal" nature of these evidences, makes these questions seem so elementary. Why? Because for the past seventy years, we have only been concerned with the "maximized" production of cultivated plants. Yield has blinded us to the biological functions of soils and soil fertility in general. Insightful outsiders of the first agricultural revolution or pioneers of the second wave had already, since the post-war period, emphasized the importance of the major ecological or mineral balances of the soil. These include researchers and farmers like Allan Savory, Masanobu Fukuoka, André Voisin, Hans Peter Rusch, Francis Chaboussou, William Albrecht, Kathy Voth, Neal Kinsey... and many others. These voices have been too little heard so far. The reflection proposed in the book you hold in your hands is also the result of fruitful reading of these authors whose agronomic currents have been unjustly forgotten in favor of the illusion of increasing and sustainable productivity. Urgency has overridden what's important!

In a society increasingly hysterical about all social issues, agriculture and ecology are not spared. Too often, faced with horror, we close our eyes and shout, "Enough!", demanding measures to solve the problem. However, some problems remain without a solution... Or at least, without a single solution, as the French writer and philosopher Éric-Emmanuel

---

[2]Holistic: pertaining to an approach that considers a subject in all its dimensions, taking into account the interactions between the various parts that constitute it, shape it, and influence it.

Schmitt perfectly analyzes and theorizes. Thus, we understand the need to halt climate change, but we do not want to stop growth. We demand healthy, residue-free, traceable, and nutritious food, but we want low food prices.

We condemn phytosanitary products, but we want food supply security...

This is what we call the tragic: the opposition of two forces that do not agree and yet both have a point. Faced with the tragedy of situations, the temptation is to prefer the drama with its elementary scheme: the good and the bad, and to propose a single solution even if it destroys the other, its reasoning, its work, its inner ecology, its “raison d'être”. Faced with the easy temptation of dramatization and hysteria around agricultural issues, we here make the bet to "tame the tragic", to reconcile ecology and economy, theory and practice, fertility and yield, productivity and soil health, abundant food and human health...

In the current context, the common sense of farmers, which is their most solid asset, is sometimes undermined. More than ever, there is a strong temptation to manage land through excess. Conventional agriculture has taken a turn that we are now rightly measuring the negative effects of. A correction of course is necessary, but upon careful consideration, and if we are not careful, other accidents are being prepared. Between two opposite excesses, there is often a middle way called "mediety," an Aristotelian concept. This middle ground is, according to Aristotle, where all virtue lies. Thus, generosity lies between avarice and prodigality. Courage is the median between fear and recklessness.The great ambition of this book is to attempt to offer farmers, regardless of their productions, cropping systems, or specifications, this middle and virtuous path that is obscured by the grand dramatizing theories of our early and last century. To achieve this, it invites you in the first parts to reacquaint yourself with the "tragic" by dissecting the excesses, the exaggerations, and the limits of conventional agriculture, as well as the exaggerated promises and the illusory successes of various forms of so-called sustainable agriculture. Then, it proposes - by restoring the rightful place of humans in the cultivated ecosystem - a new paradigm of bio-inspired agriculture, attentive to facts and the complex mechanisms and relationships of life. Finally, the last parts of this book provide a method based on a thorough understanding of soils to restore the major mineral, physical, and biological balances. Placing the soil at the center of the system thus represents a true mental revolution that it is desirable to implement today. This is why we invite the reader not to skip the reading of the first parts, which may seem overly theoretical. Otherwise, the practical implementation that we propose may appear unfounded. However, all the practice presented here is deeply rooted in a science often battered, misdirected, and misunderstood. This science is called ecology.

With this book, the author offers farmers and agronomists a condensed version of his intellectual journey and approach. Through this work, his intention is to broaden the "realm of possibilities" for all those who are willing to share his reflections and envision extensions.

INTRODUCTION

Wheat Ears of Gold

"April: 22 days of rain, 99.6 mm of cumulative precipitation. May: 24 days of rain and 195 mm. June: 16 days of rain and 120.5 mm." Personal excerpt from a farming journal, Nièvre, 1994

The year 1994 is one of those years that farmers want to forget in their lifetime, one that validates the old saying: "a good year for hay, a bad year for everything else." This excessive rainfall - although favorable for grass growth - was indeed completely disastrous for wheat grain filling.

Despite a rather mild autumn and winter in 1993-1994, which produced lush, thick stands of wheat and barley coming out of winter, the late spring weather turned promising crops increasingly dull day by day, eventually turning them to dirty yellows - grayish, then rusty hues appearing over time. By June, hopes definitively vanished with the incessant rains. In July, the guillotine fell under the combine harvester's blade. The concrete results confirmed the fears. Acre after acre, the catastrophe was confirmed. The number of grains was there, but they were thin; specific weight and thousand grain weight were miserable. The grains were hollow and wrinkled, with yields peaking at 4.5 tons per hectare (t/ha). Not enough to cover proportional and fixed costs. Let's quickly turn the page! After all, it is the fate of the farmer to navigate through time, to travel between the hope of a bountiful harvest and disappointment or happy surprise, and to rewrite a new page each autumn, each spring, with the same enthusiasm. Obviously, since each time we give the best of what we know, we are further enriched by past experiences.

However, despite the disappointment, this year 1994 would prove to be rich in lessons. In retrospect, it can be said that it was worth its weight in gold and that the poor harvests were probably the price to pay for learning. Sometimes only extreme situations provide sufficient contrasts to reveal what mere nuances don't. Thus, at the turn of a curve in the field, illuminated by the combine harvester's headlights, in the sea of grayish ears, a golden spot stood out more and more clearly. On the spot, the observation is even more surprising. Ears of beautiful yellow color perfectly hang from their stalks! Outside, the machine's bucket elevator spews out an uninterrupted flow of plump wheat, visibly heavy, of a beautiful crusty bread color. Immediately, the decision is made to bypass this spot and return quickly the next morning to see it in full daylight.

In the sunlight, on foot, eyes at the level of the wheat, in the fresh air of the early morning, it turns out that there are several places on the edge of the field where the phenomenon observed the day before is expressed. In one place, the phenomenon is even more astonishing. A main branch of a border tree, broken in the winter by violent winds and left in place, had required the sprayer to bypass it in the spring. This part of the crop had therefore not received any fungicide. Yet the wheat there was much healthier than everywhere else, and the yield must have been at least 50% higher. Upon closer inspection, this phenomenon was very pronounced near large trees, especially near oak trees, but also near some other species such as walnut and wild cherry trees. What could have happened that year?

The exceptionally abundant rainfall had reduced the impact of competition for water between the trees and the grass. Water was no longer a limiting factor that year. Only the positive effects of this cohabitation remained. It became probable that several collaborative, regulatory, ecosystemic, symbiotic, and adaptive mechanisms of Nature, known through the reading of scientific publications, had fully expressed themselves that year on the edge of the woods. Virtuality became reality. In the light of this experience and others, whether endured or caused by experimentation on the farm, the idea had to increasingly impose itself: that a sick wheat crop is not necessarily due to a "fungicide deficiency" or to fate. It is most often a wheat crop that has not benefited from the solidarity of the soil and its microbiological environment.

SOLIDARITY IN SOILS

The phenomena underlying these observations are evidently both underground and aerial. The mechanisms of nutrient transfer are now well understood. Endomycorrhizae can establish themselves for plants of different species. It is therefore possible for transfers to occur not only between plants of the same species but also between plants of different species in a very efficient manner (work by Suzanne Simard, University of British Columbia). During the damp and dim spring of 1994, the failing photosynthesis of wheat was certainly compensated by the still effective photosynthesis of large trees, this compensation mechanism occurring through the efficient exchange and transfer system of carbohydrates (i.e., sugars) in the soil. It has been demonstrated that such transfers occur between many plants, such as between ribwort plantain and sheep's fescue (Francis & Read, 1984). This mechanism is particularly pronounced when one of the plants suffers from a lack of light, which penalizes sugar production through photosynthesis. The phenomenon can be multiplied by six in the case of total shading, as in the heart of a forest. This explains the survival of young plants in forests with almost total darkness caused by mature trees. Generations of older plants collaboratively provide nutrients to younger ones, thereby ensuring the survival and growth of subsequent generations, even if they are not of the same species. In return, they benefit from "barrier," "buffer," or "collaborative" services made possible by maintaining a rich and balanced ecosystem.

NATURE'S INTERNET

Today, it is well understood that there are mechanisms for wealth distribution among plants, and that these mechanisms are facilitated by soil life. Such mechanisms have been notably highlighted by researcher Suzanne W. Simard. By using isotopic forms of phosphorus, which she spread over a 1 m^2 forest plot, she demonstrated that the elements had circulated over an area of 1,000 m^2 within thirty days. Capillary mechanisms alone are not sufficient to explain this extraordinary mobility of an element that is typically very immobile in the soil. It appears that only large trees, likely supported by mycorrhizae, were capable of exchanging this mineral and circulating it within the ecosystem. This has implications for soils: a soil that functions well is also a soil with little heterogeneity. Carbon isotopes have been used to demonstrate that two species in Canadian forests, Douglas fir and birch, exchange molecules through their roots. Trees in a forest are social beings, and communication between them is beneficial. And plants exchange much more than just nutrients. They are capable of exchanging information. A plant attacked by a pest or afflicted by a disease can release signals into its root environment that will be interpreted by others. It has been shown that in some cases, these healthy plants respond to these signals by adapting their metabolism as if preparing for impending danger. Researcher Song

Yuan (2015) from the Laboratory of Ecological Agriculture in Guangzhou planted tomatoes in pairs in 2010, then subjected the leaves of one partner from each pair to pest attack. In the presence of the root fungus, the healthy tomato begins to produce defense enzymes, typically synthesized during aggression. Conversely, if the mycorrhiza is absent, the defenses of the healthy tomato are not mobilized.

The transmission channel for both nutrients and information would be effectively ensured by mycorrhizal networks. Countless species are completely interconnected by links other than trophic chains (feeding). Some of these species cannot actually live without the exchanges they maintain with others. For instance, certain strains of microorganisms cannot be multiplied in isolated cultures.

SOVEREIGN EXCHANGES IN FIELDS

In the case of our wheat field in 1994, the "photovoltaic panels" of the trees evidently displayed solidarity with the crop. The hypothesis is that the equalization fund, the solidarity economy put into action, allowed perennial plants with a wide canopy fully exposed to the sun to supply short-cycle herbaceous plants, whose development was concentrated mainly over a few months, unfortunately without much sunlight in that year, 1994. Following this thread on the family farm with mixed farming, it becomes evident that these biological solidarity mechanisms are not exclusive to forests and trees. Cultivated ecosystems—when biology is both active and well-structured among its major families of microorganisms—are always the site of intensive mechanisms for redistributing nutrient wealth. The heterogeneity of a plot is always indicative of a lack of solidarity among plants, an absence of a distribution system, and therefore a microbiology that needs to be restored.

First conclusion: we must shift the focus of our attention towards soil function and soil life. 80% of aerial issues are linked to its functioning. Each farming intervention has a primary objective, but also known or unknown, intentional or unintentional peripheral effects. Should all operations—soil cultivation, fertilization, soil pathogen control—therefore be globally reasoned to weigh the benefits/risks balance as comprehensively as possible? To carry this coherence through to the end, shouldn't every farming action take into account its impact on soil vitality?

A LEAF MICROBIOTA

The second communication and collaboration pathway through living organisms is aerial. The phyllosphere alone constitutes an extremely rich ecosystem. When buds open, the leaf surface is practically sterile, but will quickly be colonized by countless microorganisms. As it unfolds, bacteria, fungi, and yeasts will be brought by rain and wind. The most abundant belong to the yeast family and are particularly adapted to resist sunlight by producing melanin. There are also numerous non-pathogenic fungi such as Aureobasidium pullulans and Cladosporium, which complete their entire life cycle only on leaves. They are equipped to withstand the unique environment of leaf surfaces: resistance to the deadly UV rays of the sun through melanization of their hyphae, which also protects them from intense dehydration and bacterial attacks. They primarily feed on leaf exudates of the host plant, consisting of amino acids, sugars, and inorganic ions essential for their growth, thus passively defending against pathogenic fungi through food competition. Another example of interaction: the conidia of *Botrytis cinerea* (a cryptogamic disease), present on a leaf, exude substances after dew or rain moistens them, containing enough amino acids and sugars for bacteria to proliferate and prevent the germination of these same conidia. Thus, the disease itself feeds the "police" responsible for controlling it.

The older a tree is, the more diversified fauna and microbiology it possesses, biodiversity to ensure its own protection and that of its neighbors. So, could wheat, the newcomer of the year, have benefited from the benevolent protection of the "arsenal" of its older sibling? Could destroying non-pathogenic or even beneficial species on a plant through fungicides leave it completely defenseless instead of ensuring protection?

Second conclusion: Fungicides, herbicides, biostimulants, foliar fertilizers necessarily have an influence on the phyllospheric flora. Shouldn't they therefore be reasoned, designed, and managed with this perspective in mind ?

INCREDIBLE RESILIENCE

These initial concluding questions could only lead to an ongoing quest. Since then, answers have been provided, but new questions have arisen: **Are we not paying too much attention to what is potentially toxic (pests, diseases, etc.) and insufficient attention to what builds the health of**

plants, animals, and humans (fertility in the broad sense)? One of the most striking examples that led us to ask this politically incorrect question is the Chernobyl disaster that occurred on April 26, 1986. At the time, some environmentalists predicted the emergence of a lunar-like landscape with the disappearance of most of the fauna. During the event itself, there were indeed massive losses in fauna and flora, besides the tens of thousands of people who died directly or indirectly. Thirty years later, the forest now covers 70% of the affected area, doubling its surface. "At present, it is difficult to find traces of harmful effects of radiation on biota in the immediate vicinity of the radiation source (a few kilometers from the damaged reactor), and in the rest of the territory, plants and wild animals thrive due to the elimination of the main natural stress factor: humans," concluded scientists from the International Atomic Energy Agency in their report (IAEA, 2006). There seems to be another complementary hypothesis: Plants subjected to significant abiotic stress produce more antioxidants, polyphenols, tannins, tocopherols, etc., and allow the entire food chain to benefit from this abnormally, but fortunately, high concentration of protective molecules (Jameel, Q. Y., & Mohammed, N. K. 2021). The worst is never certain, and today we have proof that no fertility is definitively lost. This observation raises an essential point for our society, which is so concerned with the precautionary principle: **health is as much a matter of the presence of vital principles as the absence of toxic substances.**

SOILS ARE NOT DEAD

When we imagine that lifeforms endure extreme conditions, such as the cores of nuclear reactors (Deinococcus radiodurans withstands radiation levels of 15,000 to 30,000 Gray. For reference, the lethal dose for humans is 10 Gray), deep-sea trenches (chemolithotrophic archaea thriving without the aid of sunlight or organic debris at pressures of 1,000 bars), the most aggressive acidities (Picrophilus oshimae withstands pH 0), and extreme temperatures (some archaea thrive at temperatures up to 120°C) near "black smokers" or in geysers. When we think of life forms like the tardigrade, a cousin of arthropods, approximately 1 mm in size, which colonizes every corner of the planet and all environments under the most challenging conditions: exposed for ten days to the vacuum of space and ultraviolet radiation 1,000 times greater than on Earth, this tiny animal survived by repairing its own DNA damaged by cosmic rays. It can endure a tremendous thermal range, from -270°C to +150°C, vacuum, or pressures of 600 bars. It can stay without water or food for more than ten years! Dehydrate to 3% water content, then "rebirth" once rehydrated. Undergo freezing and come back to life once thawed... even after being frozen for 2,000 years in ice. So, yes, when we become aware of all these incredible manifestations of life, so far removed from the requirements of human metabolism, we are forced to come to the obvious conclusion that soils contain treasures of self-repair and infinite possibilities to rebuild evolutionary strata damaged by aggressive agricultural practices. Cultivated soils are highly resilient with astonishing rapid repair capabilities by following some simple rules. What gives them such self-repair capabilities is the tremendous reservoir of biodiversity and productivity they contain, still largely unknown and therefore underexploited.

Considering damaged, polluted, eroded soils as "dead" is anthropocentric. No, they are not, but we more or less consciously sense that an abused soil, a scratched planet, a degraded physicochemical environment will only be habitable for "pioneer" creatures endowed with adaptive qualities that humans do not possess. Humans will remain biologically alive only if the Earth and soil continue to maintain temperatures, pH levels, air quality, and food nutrient density within the narrow ranges that allow survival. Hence, our conviction that humans are alive when the soil is alive. Moreover, the Homo economicus, which is the farmer, will also survive only if his primary tool, the soil, retains sufficient vitality to provide free and unsuspected functions: water purification, regular plant nutrition, crop health, carbon storage. The soil is holistic, and none of these functions can be properly ensured alone. It's all or nothing. Innovation in this regard is to approach the soil from a global perspective and adopt strategies based on the whole.

A

Building an Agricultural "Renaissance"

"Finally, we admit that the sun itself occupies the center of the world. All these things are taught to us by the law of order in which they follow one another, and by the harmony of the world, provided only that we look at things themselves with both eyes, as it were."

Nicolaus Copernicus, *De revolutionibus orbium coelestium*, 1543

The Copernican revolution allowed us to move away from a geocentric view of the world to integrate the reality of heliocentrism. The "whole" is not the Earth system. The latter is included in the solar system, itself included in the universe. This shift in perspective deeply shook society during the Renaissance. We believe that the necessary transition from an agronomic science focused on yield to an approach centered around soil fertility constitutes a transformation of a similar magnitude. The system of representing crop performance, the role of the farmer, and the place of humanity in the agrosystem are all poised to undergo significant evolution. For agriculture to undergo its agroecological transition, we not only need new techniques but also must change our mindset and achieve a second Green Revolution.

1 • KNOWLEDGE SEGMENTATION

"It is a sad thing to think that nature speaks and mankind does not listen."

Victor Hugo, *Carnets*, 1870

Modern agricultural science has been built on the segmentation of knowledge and fields of application. However, an ecosystem like the cultivated ecosystem constitutes a whole that cannot be reduced to the sum of its parts.

1.1 Strengths and weaknesses of the segmentation of science

1.1.1 A fruitful specialization of human activities

The only truly vital basic activities for human beings are drinking, eating, clothing oneself, and protecting oneself from the elements. Throughout thousands of generations, these survival activities, essential for the reproduction of the human species, occupied the majority of time. When humans appeared on Earth, all production and consumption functions were carried out by the same individual: the one who hunted, fished, or gathered also consumed the food collected, either by themselves or by their immediate relatives. This state of affairs persisted for hundreds of thousands of years. Later emerged the first organized communities, developing internal relationships slightly more complex than simple clans or tribes. This complexity, tens of thousands of years ago, was accompanied by specialization, evident in the remnants we find today in Neolithic or Paleolithic sites: tools, adornments, paintings, sculptures, cave dwellings, and more. A beginning of stratification within human society led, based on individual qualities, physical strength, the need for social recognition, and ambitions, to a division of labor. Leaders, artists, priests, soldiers, and farmers carried out very specific tasks; skills requiring learning and transmission developed; growing sophistication and efficiency allowed humans to inhabit inhospitable environments; and they also increased their population to unimaginable proportions while reducing the threat of famine. Maslow, who proposed the hierarchy of needs, also provided a shortcut to the evolution of human societies: after meeting basic needs, the necessity to fulfill more "noble" requirements quickly arises—personal development, group recognition, aesthetic asceticism, and even spiritual quests. This progression is only possible thanks to securing basic supplies, leading to serenity and time to engage in other pursuits. Our ancestors only achieved this comfort through the rationalization of all processes of societal life. Recent developments sometimes had excessive consequences on individuals: Taylorism, Stakhanovism, etc. However, these openings were undeniably effective; these excesses quickly found gentler forms, taking into account human resources: working hours, strenuousness, recognition, task evolution, mobility, and training.

1.1.2 Unfortunate disconnections

One of the less immediately visible consequences of this slow evolution is the disconnection between individuals, processes, and activities. There are relationships and information, but less communication. We will explore the significant impacts this growth crisis has on our perception of reality and the potential for progress that lies ahead. Our society has developed muscles and skeletons, but the nervous system, the lymphatic system with all its information and regulatory substances, have not kept pace. When the hunter, producer, consumes their own products, they quickly make the connection between their practices and the organoleptic qualities of what they consume; the farmer knows this, they "feel" it. How many of us treat what we know will be consumed by ourselves or our loved ones differently? Every time the consumption circle widens, anonymity and irresponsibility increase. We have gone from the family circle to the village, from the district to the region, and now from the country to the world. We have shifted from a universe of micro-societies in restricted territories, practically autonomous and producing 90% of their consumption, to an archipelago of highly specialized units supplying the global market: how can the dissatisfied consumer's regulation system work? Even if markers "track" products, the self-regulation system remains clogged and slow: legal barriers, administrative delays, and simply distances isolate the consumer with their needs and complaints from the producer.

The second factor that can hinder self-regulation and the exercise of common sense is that ultra-specialization generates populations of researchers and technicians who use a language inaccessible to the majority of fellow citizens. The elderly, non-specialists, and operational individuals are quickly relegated to the role of spectators. Claude Michelet, in "Des grives aux loups," describes this memorable episode of the arrival of a tractor on a farm: instinctively, the technician turns to the son to explain how the machine works. The "elderly" person is set aside, with the assumption being that the development of new concepts, techniques, and machines inevitably disqualifies someone who has been nurtured by a culture from another era. We are witnessing the marginalization of older individuals, neglecting the fact that the emergence of new tools does not change the fundamentals: the functioning of soil, agronomy, and plant physiology remain constant. Neglecting acquired experience is a monstrous waste of knowledge, a "bonfire" of living libraries, and is undoubtedly the main reason for agricultural practices' deviations. These deviations have led to a neglect of "good stewardship" management of soil heritage: neglected fallowing, plowing at arbitrary depths, abandonment of legumes and cover crops, economic rather than agronomic crop rotations have gradually led to declines in organic matter, increased erosion, and increased compaction. The power of mechanical and chemical tools has led to the belief that ancestral knowledge can be bypassed. **True innovation is incremental and builds on past knowledge; it does not negate it.**

Finally, agricultural production has largely been following the orders of consumers. For years, consumers have given clear indications to the agri-food sectors to reduce the cost of food. They have also imposed their preference for perfectly sized and defect-free fruits and vegetables. Every time a consumer has chosen the cheapest product, every time they have rejected a blemished or misshapen fruit, they have given extremely clear orders to producers, via the chain of department managers, central buyers, wholesalers, and cooperatives. Even today, the outward appearance of fruits and vegetables is often the number one selection criterion. The pact between farmers and consumers is certainly in need of revision and, let's be optimistic, is on the verge of being revised by initiatives such as short supply chains and communication from supermarkets, which are increasingly focusing on and promoting regional sourcing with fair remuneration for farmers.

1.1.3 A growing awareness

The "Green Revolution" aimed at one objective: increasing yields. This has occurred in many situations. Humans have developed increasingly efficient agricultural systems that can feed a growing global population. Herbicides like glyphosate have enabled the cultivation of two crops annually, or even three in certain climates. Conservation practices such as simplified cultivation techniques and no-till farming owe much to the use of this same molecule. Malthusian predictions (which were mistaken for reasoning at a constant technological level) have not been borne out. The objectives of the Treaty of Rome, signed on March 25, 1957, aiming for the creation of an agricultural Europe, have been achieved. Europe, including France, has become more than self-sufficient and is even an exporter. The famines that persist today are more due to political problems and poor distribution of wealth and food than to an inability of the planet to produce enough food.

This increased performance, resulting from a high level of technicality and advanced specialization, is reaching its limits. Most soils have a natural fertility lower than it was a century ago: with the same variety, same fertilizers, and same soil management practices, yields today would be lower than those of the early twentieth century. We are beginning to perceive that the increasingly significant inputs of mineral and organic fertilizers are compensating less and less for the loss of life in the soils. This losing battle will be limited by economic possibilities. As we will see, "doing without" the free contribution of an effective food chain in the soil is extremely costly. We are certainly not very far from this point of unprofitability.

Conflicting demands

> Keeping prices low or paying more attention to animal welfare? Preserving the appearance of the product or conserving the environment? Agriculture finds itself trapped between conflicting demands. In the frameworks that some have attempted to theorize, such as Gregory Bateson (Palo Alto school), no honorable way out is possible because respecting the first part of the equation leads to neglecting the second. The demands of today's consumers therefore seem impossible to satisfy, unless these unattainable demands conceal another issue, a more nuanced and therefore more difficult to express demand: to re-establish dialogue and better understand what is on their plate? To create an agriculture of life and nature? These two proposals, on the other hand, are achievable, sustainable, and desirable.

1.1.4 Building a new model that respects today's knowledge

Today, the harmful effects of certain conventional agricultural practices are well known. However, it would be unfair to discredit past agriculture with today's understanding. Agriculture as we have known it so far has fulfilled its role, and it is precisely because we are no longer hungry today, because we have achieved food security, that we can now comfortably consider taking different and more virtuous paths in terms of ecology and humanity.

The legacy of this agriculture should not be entirely cast aside. Reaching its limits, however, does not mean that we should reject all developments in agrochemistry of the past fifty years. Let us not make the mistake of completely rejecting a tired world in favor of another that appears radiant but is already known not to keep its promises. Let us not repeat the error we made seventy years ago by rejecting agronomic rules built upon the experience of generations of farmers and breeders, in favor of an agriculture finally rid of weeds, pests, and diseases by chemistry, or so we thought. The best way to navigate over a long distance is certainly not to make large turns to the right and then correct with the same amplitude to the left. It is primarily about looking in the direction of the goal we wish to achieve. Here, we will provide the pathways that transitional agriculture can take toward a new art that is under construction. In this evolutionary phase, we will attempt to identify the most limiting factors, making sure not to confuse the **main objective, which is to revitalize the soils**, with the means, which are broad. The means, mere tools, include soil management, plant species, genetics, fertilization, and cover crops. "The wise man points to the moon and the fool looks at the finger." To paraphrase this Chinese proverb, it is not about focusing solely on the finger, which are the cover crops and no-till practices. **Our moon is the living soil.**

1.2 Limits of agricultural segmentation :

1.2.1 Science has divided the whole into individual parts

Modern science owes much to René Descartes and his work *Discourse on the Method of Rightly Conducting One's Reason and of Seeking Truth in the Sciences* (1637). In this work, the philosopher provides various rules for the proper conduct of thought. After praising doubt "for each thing", he adopts the principle of "dividing each difficulty into as many parts as possible, and as many as are necessary, in order to solve them better". This principle of division is the foundation on which modern science has been built. However, Descartes is quick to emphasize the need for synthesis in his Method of Thought, which he details by "making enumerations everywhere so complete and reviews so general, that I would be assured of omitting nothing". It has to be said, however, that modern science does not sufficiently undertake this essential work of bringing together interconnected knowledge and disciplines. In so doing, our societies have certainly been able to develop ultra-high-performance technologies and extremely advanced knowledge in specialized fields. But they have shown little capacity to understand the complex phenomena of ecosystems.

1.2.2 An example: systematic overfertilization of wheat after wheat

Current recommendations for fertilizing wheat provide a clear example, among many others, of how farmers and, more broadly, cultivated ecosystems can lose out due to the compartmentalization of research. The technical recommendations for phosphorus and potassium fertilization for wheat crops following another wheat crop (known as wheat-on-wheat) are regularly higher than those for wheat following a fallow (the first wheat). However, in theory, there is no reason for the same wheat crop to have higher nutrient requirements. At the same yield and variety, the theoretical needs are the same. So why is the cultivated ecosystem in this specific case losing out so much in terms of nutrient mobilization? When questioned about this, the "plant nutrition" teams of the technical organization responsible for these recommendations simply respond that from a yield perspective, trials have shown "statistically significant positive responses to higher fertilizer applications." However, the sole indicator of "yield" does not explain the underlying mechanisms.

■ A "take-all effect" leading to enhanced fertilization of wheat following wheat crops

To find the root of the problem highlighted here, it is necessary to look at a broader issue related to the practice of wheat-on-wheat. Indeed, this practice often leads to the harmful establishment of a cereal disease called "take-all" caused by a fungus (*Gaeumannomyces graminis*). Take-all disease limits the root development of the cultivated plant and thus its ability to obtain minerals. Therefore, it becomes clearer how systematic overfertilization in the context of wheat-on-wheat improves yield.

■ Links between take-all disease and redox

Plant pathology, microbiology, and other scientific disciplines are essential for a comprehensive understanding of the specific issue of fertilizing wheat-on-wheat crops.

In the early twentieth century, the French scientist Albert Demolon (in *Dynamique du sol* [*Soil Dynamics*], Dunod, 1932) already reported how the action of take-all disease ceases when organic fertilizers are applied. Studies have shown that the susceptibility of cereals to take-all disease is often linked to the unavailability of manganese in the soil[3].

---

[3] Timonin, 1972 ; Arnott et al., 1991 Heckman et al., 2003, Mathre, 1995, Singleton et al., 1992.

Based on current knowledge, it appears that the plant's level of nutrition, specifically having sufficient manganese, confers resistance to the disease. The context of wheat following wheat crops favors take-all disease because the preceding wheat creates conditions unfavorable for the bioavailability of manganese.

It is worth noting that applications of oxidized manganese have been found to have no effect on improving yields in the presence of take-all disease inoculum. As for foliar applications of manganese, regardless of their form (reduced or oxidized), they have never been able to improve yields in the context of confirmed take-all disease. Therefore, the positive action of manganese may not solely occur through direct assimilation by the cultivated plant. It seems that the beneficial effect of manganese on the disease involves root assimilation in reduced form following biochemical reactions in the soil. This hypothesis has been confirmed by other experiments demonstrating the effectiveness of inoculating soil bacteria capable of reducing manganese in combating the disease[4]. Hence, it is better understood why the applications of manure by A. Demolon in the early 20th century reduced the impact of take-all disease. These applications likely derived their fungistatic effect from the inoculation or stimulation of manganese-reducing bacteria. The provision of reduced and bioavailable manganese by soil bacteria constitutes a natural antibiosis mechanism that helps regulate the expression of the disease.

In the short term, applying manganese chloride to the soil or seed has proven effective. Other practices can also mitigate the negative effects of take-all disease: rotations with species such as oats (whose rhizosphere chemically reduces manganese), incorporating organic inputs of animal origin, or applying *Bacillus mucilaginosus* to the soil or seed, which has the ability to reduce manganese and make it available. A sufficient level of nutrition can be provided for free by soil biological chains when they are active. These methods remain the most effective and economical.

The virtuous mechanism of plant protection is a specific case of phenomena formerly well known as "soil disease suppression." There are hundreds of other such mechanisms. In fact, regarding the resistance of the soil itself, we now understand that these mechanisms involve complex

[4] P. Marschner, J. S. Ascher & R. D. Graham, Effect of manganese-reducing rhizosphere bacteria on the growth of Gaeumannomyces graminis var. tritici and on manganese uptake by wheat (Triticum aestivum L.), Biology and Fertility of Soils volume 12, pages33–38, 1991

interactions of microorganisms, and it is these complex interactions that constitute these "nutrition and resistance factors."

1.2.3 Conclusion

What conclusion can be drawn from all of this? Agronomy has not escaped the hyperspecialization of research departments. Consequently, all the complex phenomena related to living organisms and involving multiple factors are often ignored and rarely studied in their entirety. Successes in productivity are seldom analyzed to identify their causes, leading to modeling soil/plant relationships based on an equation: one problem equals one solution. And never mind the side effects, never mind multifunctionality, never mind complex interactions, never mind living organisms.

2 • RE-ESTABLISHING THE LINK BETWEEN FERTILITY AND PLANT HEALTH

The division of agronomy into distinct specialties has been partly detrimental to agriculture. There are two areas in particular that deserve to be closely connected: plant protection and health on one hand, and soil fertility on the other. Here, we will see how these connections are established and how they form the basis of a new approach to soils. We believe that at least 80% of plant aerial issues are related to soil functioning.

2.1 "Cides" have an impact on natural soil resistance

In his book *Healthy Crops: A New Agricultural Revolution* (1985), Francis Chaboussou, an honorary research director at INRA for nearly fifty years, starts from the observation, with dozens of examples, that "natural resistances of soils" of microbiological origin are destroyed by the implementation of pesticide protections.

Indeed, the massive use of agrochemicals has had unintended side effects on soil life and living mechanisms that ensured "total incompatibility between the pathogen and the soil-plant system." Agrocenoses[5] that had never allowed a pathogen to express itself become sensitive to it. Francis Chaboussou relies on his demonstrations notably on the documented case of vascular fusariosis of melon. The disease eventually established itself in plots previously completely free of it in the lower Durance valley following the use of fumigation programs or anti-"corky-root" thermal treatments. Knowing that soil biology has a significant implication in these phenomena of natural resistance, it is well understood how biocides or other treatments destructive to life can suppress the installation of such mechanisms. Furthermore, we often talk about "imported" diseases. Sometimes, it is indeed the importation of an exotic pest, particularly well adapted and for which the soil system has not had the opportunity over the centuries to develop effective regulatory mechanisms. But often, it is

[5] an artificial ecosystem created by humans including cultivated plants, weeds, and associated parasitism

not the importation of the pest that is the cause of its establishment. It is ecosystemic barriers in the soil that have eventually been overcome under the repeated action of biocides or other agricultural practices.

2.2 "Cides" sometimes have transient "positive" effects in soils

We have just seen that biocides regularly cause biological dysfunctions in soils. For example, their action can suppress a soil's ability to make plants resistant to disease or pests. Sometimes, biocides, depending on the molecules and methods used, are, on the contrary, very beneficial **in the short term** to the physiology of cultivated plants and their productivity. The use of insecticides, now banned in France, such as metam-sodium, carbonyl sulfides, carbofuran, or chlorpyrifos, often accompanies an immediate "boost" effect on vegetative growth. They seem to act like fertilizer. How could this observed resurgence in vigor be solely linked to the direct action of chemical molecules? A more realistic hypothesis is to consider that the collateral effect of destroying an entire soil life releases abundant amounts of nutrients. Positive effects then appear due to a better level of plant nutrition ensured by the use of the "cide" which releases the nutrients contained in the destroyed living organisms.

2.2.1 "Cides" create transient fertility

Significant increases in agricultural yields can be achieved by destroying most forms of life present in the soil and reducing organic matter levels. Indeed, all the mineral and organic materials constituting the building blocks of these living or biologically derived materials become available to plants when they are eliminated. Therefore, it can be asserted that, in many cases, the observation of increased yields in randomized experimental designs with a solid statistical basis has positively endorsed certain technical practices, even though they were only the most "efficient" at destroying some of the soil life and thus making more nutrients available to plants. Hence, the widespread impoverishment of soil biodiversity and the subsequent decrease in organic matter levels ensued.

2.2.2 Decapitalization at work

All products ending in "-cide" necessarily represent a setback in agronomic terms. The observed effectiveness of a biocide in fields simply signifies that one of the regulatory links has failed, and its regulatory function has been temporarily supplanted by a product invented by humans. Clearly, this is a forward escape that foreshadows the cascading collapse of more and more vital links in the chain of life. Resistances emerge, failures multiply, deadlocks become increasingly common. It is a battle lost in advance by humanity. The more artificial the growing environment becomes, the further it deviates from the optimum, and the more invasive and extremophile species find themselves in competitive conditions favorable to them (pathogens, weeds, etc.).

2.2.3 Conclusion

All too often, modern agriculture has achieved its yield successes on the ashes of soil life, depleting its capital reservoir. The assessment of unintended effects has been poorly conducted. Agronomist Francis Chaboussou, in his book *Les Plantes Malades des Pesticides* (*Plants Sickened by Pesticides*) (2011), was a pioneer in identifying some of these mechanisms. The book *Recueil des Effets Non Intentionnels des Produits Phytosanitaires* (*Collection of Unintended Effects of Plant Protection Product*) (2002) also shows that research has paid little attention to unintended effects on soil microbiology and fertility. No chapter is dedicated to this issue. This represents a significant scientific bias.

2.3 Trophobiosis: plants programmed to be sensitive or resistant

2.3.1 A new perspective on pest outbreaks

Francis Chaboussou's research has shown that we can go even further in establishing links between soil fertility and plant health. Let's take the example of aphid outbreaks. The scientific community generally admits that aphid movements and outbreaks (or those of other insects classified as crop pests, such as the Colorado potato beetle, for example) are random, unpredictable, and inexplicable given current knowledge. Applied research has therefore largely abandoned the idea of finding the cause of these phenomena and instead focuses on destroying the pest without seeking to identify the cause of its proliferation.

- **The pests "choose" their target based on their nutritional value.**

However, the question of the complex relationships that crop pests have with their targets has already been addressed since the 1970s by Francis Chaboussou. He established in his theory of trophobiosis[6] that "every vital

[6] Trophobiosis: trophic relationship between two species where one provides food to the

process depends on satisfying the needs of the living organism, whether it is a virus, plant, or animal" and that correspondingly, "every parasite becomes virulent only if it finds in the plant the nutritional elements it needs." Plant organs will only be attacked "to the extent that their biochemical state corresponds to the trophic requirements of the parasite in question."
The most commonly accepted explanations for the spread of parasitism (such as that of aphids mentioned above) invoke the presence of harmful organisms in the environment or the absence of regulatory elements, notably predators (hence the idea of resorting to helpers like ladybugs). However, the phenomenon of trophobiosis shows us that there is primarily a true determinism of the plant to be affected by parasitism and that this determinism comes from the physiological state of the plant, which is itself dependent on the effectiveness of trophic chains and fertility in the soil. Pest outbreaks are linked to the establishment of trophic conditions that are extremely favorable to them.

- **Parasites, "cleaners" of the ecosystem.**

The predator that is the insect carries out the work for which ecosystems have selected it over thousands of years of evolution: cleaning and consuming weak, nutrient-poor, and uncompetitive plants that are marked as "without a future" within the ecosystem. Philip Callahan, the American entomologist who originated this theory, has demonstrated the reality of this phenomenon through compelling experiments. To achieve this, he relied on the fact that the infrared signature of plants evolves based on their sugar and mineral content. Insects are equipped with vision that allows them to detect plants over long distances based on their infrared signature. Signals emitted by "weak" plants are detectable up to several kilometers away. By placing a malnourished plant with low Brix (low sugar content) behind an ordinary glass panel (which does not allow infrared rays to pass through) and on the other side behind a treated glass panel allowing infrared rays to pass through, the following phenomenon occurs: insects rush to the treated glass surface (allowing IR passage) in hopes of feasting. The untreated glass surface, impermeable to IR, is not visited. Insect flights and their destinations are not random; they are guided by the IR signatures of plants determined by the plant's nutritional level itself, which depends on soil life.

---

other. The theory of trophobiosis implies that the pest only attacks its target if the target itself is adapted to providing suitable food for the pest.

2.3.2 Plants protected by their proteins and complex sugars

We have seen that the nutritional status of the plant plays a crucial role in its susceptibility to parasitism. However, what is the desirable nutritional state and how can it be promoted? Thanks to Francis Chaboussou, we know that plant health is associated with high levels of protein and complex sugars produced by the plant, and also when proteosynthesis greatly exceeds proteolysis. He also suggests that soluble nitrogen is the nutritional element par excellence favoring the development of various ailments. Plants with efficient proteosynthesis and glucosynthesis are less palatable, indigestible, or even toxic to parasites. On the contrary, plants rich in soluble elements (soluble nitrogen, minerals, amino acids, reducing sugars, etc.) are highly palatable and conducive to parasite development. This is partly why certain excesses of vigor are detrimental to plants with high concentrations of soluble nitrogen in tissues, promoting the development of a wide range of parasitic diseases.

At the same time, it turns out that the efficient synthesis of proteins and complex sugars occurs in soils where plant nutrition is favorable. Soils rich in trace elements like volcanic soils are often "resistant" to pathogens. Indirectly, parasitoses result from numerous deficiencies in the soil, many of which are caused by agronomic practices (especially pesticides). This established link between soil fertility and the synthesis of complex compounds by the plant is at the heart of fertilization practices as we envision them.

Managing health through trace elements

Many farmers believe that applying trace elements to the soil is not useful, and that only foliar applications are beneficial. However, trace elements are also useful for activating the soil's toolbox through enzymes, and as precursors for effective proteosynthesis activity. If a plant requires foliar fertilization with trace elements, it is a sign that soil fertility has not performed its role effectively.

2.3.3 Sugar content as a good indicator of protection

The attacks of corn borers on corn provide evidence of this positive tropism linked to trophobiosis. Indeed, these attacks are often localized at the base of the ear. When measuring sugar levels at various points of the corn plant using a refractometer, we repeatedly observe lower sugar concentrations at the base of the ear. They are lower because the ear acts like a sucker and thus reduces the sugar gradient at the stem attachment. These conditions would be more favorable for larval development. When the sap is too concentrated in sugars, the risk of alcoholic fermentation in the larval and adult moth digestive system is increased. This creates conditions in their digestive system that are conducive to alcoholic fermentations. The sap then poses a high risk of toxicity to the predatory insect. An alternative and sustainable solution to combat pests is to create favorable nutritional conditions that allow the plant to produce saps rich enough in sugar, vitamins, and minerals. This is what we call nutriprotection. Insects, capable of visually analyzing this nutritional state of the plant, tend to move away from it.

2.3.4 Trophobiosis applied to cryptogamic diseases

The phenomenon of trophobiosis is not only valid for insect and mite pests but also for many diseases. Here is a compelling example from the fight against Esca. This cryptogamic disease attacks the wood of the vine. Its impact is extremely significant, particularly in France, where it is considered "the most worrying vine wood disease" (Lecomte et al., 2019). According to the DGAL (French General Directorate for Food, 2012), 13% of the French vineyard is affected by wood diseases. Esca is characterized by rapid vine deaths, often occurring after alternating periods of water stress followed by humid heat. The disease is caused by several species of fungi, the most documented being *Phaeomoniella chlamydospora*, *Botryosphaeria* spp, *Phaeoacremonium aleophilum* and *Fomitiporia mediterranea* (formerly *Phellinus*). The development of this fungal flora blocks the plant's vessels, causing the plant to enter a state of apoplexy, explaining the rapid onset of the disease. Chemical control methods are ineffective.

We have observed on several occasions that vines with good root levels were less susceptible to the disease. This led us to recommend to many winemakers to intensify their efforts to achieve deeper root depths. These results were achieved through the implementation of approaches described in chapters 19 and 20.

But how could actions taken on the soil improve the health of the wood? Once again, we must return to the principles of trophobiosis (preference of the pest for the target that is most nutritionally suitable). After a period of drought, the return of rain favors the mineralization of organic matter in the soil, which then becomes maximal. The absorption of a significant nutrient flux by the plant is accelerated. The sap, rich in soluble elements, then becomes a favorable culture medium for Esca. When the vine is more deeply anchored (not only in the superficial zones of the soil sensitive to mineralization), it is less exposed to variations in soluble element concentrations in the soil solution. Working on root depths is not only useful for increasing the volume explored by plants. It also affects the quality of the plant's nutrient supply, its ability to produce complex carbohydrates and proteins that protect it from external aggressions. Anything that improves the plant's nutrition also protects it against attacks by insects, mites, or fungi.

2.3.5 Brix as an indicator of plant health

Below is a tool that allows for the assessment of plant susceptibility to parasitism and potentially for managing "protective nutrition" based on the plant's proven sensitivity. The level of proteosynthesis is difficult to assess in the field. However, measuring sugar and mineral levels in plant tissues in degrees Brix is very simple and inexpensive using a refractometer.

Operating principle: Plants produce sugars through photosynthesis while simultaneously accumulating minerals. This total dissolved salts (TDS) content depends on the efficiency of photosynthesis and crop nutrition. It can be measured with a refractometer using a few drops of sap.

Objective and utility: Measure the neutral or positive impacts of crop management. Trigger only profitable inputs: the evolution of Brix, measured just before and one hour after foliar fertilization application (on a surface that can be just a few square meters), allows for quickly assessing the effectiveness of the supposed rebalancing. An increase of 0.5 to 1 point indicates a high probability of profitability for the operation. Stagnation may seriously raise doubts. Then, the decision is made to extend the operation to the entire plot or not. This technique also allows for observing the depressive effects of numerous foliar applications of phytosanitary products on photosynthesis.

Procedure: After sap extraction, take a few drops and place them on the refractometer's optic. The Brix value is directly read on the refractometer**.** Advantages: A cost-effective procedure that allows for precise management of inputs by directly soliciting the plant to determine what suits it best.

Limitations of measurements and implementation conditions: The values obtained are highly dependent on several environmental factors and can only be interpreted taking these into account: Brix levels are at their lowest in the morning following a period without photosynthesis (night) and at their highest in the late afternoon. Cloudy and cold conditions can only generate low values due to lack of light energy. When values are low for these reasons, it is then more difficult to discern significant differences between different treatments. Another scenario: when atmospheric pressure is low, plants mobilize their mobile reserves (sugars and minerals) towards the roots to preserve their ability to restart in case their aerial parts are destroyed by a violent storm (wind, hail...). Besides the raw value (see table below) depending on the quantity of sugars and minerals, the important qualitative point is the complexity, polymerization, and length of the molecules dissolved in the sap. The more complex the molecules, the more the light passing through the refractometer lenses will be diffracted. Therefore, qualitative sap will result in a milky appearance of colors in the device's optics with a blurry demarcation line between the white and blue parts. This aspect depends much more on the size of the molecules than on the calcium content, as is often claimed.

Finally, in case of high Brix values, it is important to ensure that we are not facing "abnormal" situations of sugar storage, which can have two main causes: 1) a boron deficiency, which is absolutely necessary for sugar transport 2) an inability of the plant to transpire sufficiently to ensure its cooling, leading to sugar concentration. A simple way to check this is to measure the temperature difference, during the hottest hours of the day, between the ambient air and the temperature of the foliage. This difference should not exceed 2 degrees Celsius. The measurement can be quickly performed using an IR thermometer, first on the vegetation itself, then on a vertical white paper sheet. The temperature of the latter will serve as a reference, as it is very close to the ambient air temperature.

	Very low	Low	Average	Good	Exceptional
FIELD CROPS - LEAVES/STALKS					
Alfalfa	*4*	*8*	*12*	*16*	*20*
Cock's-Foot	*2*	*4*	*6*	*8*	*10*
Corn (stalks)	*6*	*10*	*14*	*18*	*24*
Legumes	*4*	*6*	*10*	*14*	*18*
Plantain	*4*	*6*	*10*	*14*	*18*
Ryegrass	*4*	*8*	*12*	*14*	*16*
Small grains	*4*	*8*	*12*	*16*	*18*
Sorghum	*6*	*10*	*12*	*16*	*18*

	Very low	Low	Average	Good	Exceptional
VEGETABLES - VEGETATIVE PARTS					
Asparagus	*2*	*4*	*6*	*8*	*10*
Beets	*4*	*6*	*8*	*10*	*12*
Bell Peppers	*4*	*6*	*8*	*10*	*12*
Broccoli	*4*	*6*	*8*	*10*	*12*
Cabbages	*4*	*6*	*8*	*10*	*12*
Carrots	*4*	*8*	*12*	*16*	*18*
Cauliflower	*4*	*6*	*8*	*9*	*10*
Celery	*4*	*6*	*8*	*10*	*12*
Chilies	*4*	*6*	*8*	*9*	*10*
Corn (stalks)	*6*	*10*	*14*	*18*	*24*
Cucumbers	*4*	*6*	*8*	*10*	*12*
Curly Kale	*8*	*10*	*12*	*14*	*16*
Endives	*4*	*6*	*8*	*9*	*10*
Escarole	*4*	*6*	*8*	*9*	*10*
Garlic Bulb	*24*	*28*	*32*	*36*	*40*
Green Beans	*4*	*6*	*8*	*9*	*10*
Green Peas	*4*	*6*	*8*	*10*	*12*
Kohlrabi	*6*	*8*	*10*	*11*	*12*
Lettuce	*4*	*6*	*8*	*9*	*10*
Onions	4	6	8	9	10
Parsley	4	6	8	9	10
Peas	*6*	*8*	*10*	*12*	*14*
Potatoes	3	5	6	7	8
Romaine Lettuce	4	6	8	9	10
Rutabagas	4	6	8	10	12
Squashes	6	8	10	12	14
Sweet Corn	6	12	16	20	24
Sweet Potatoes	6	8	10	12	14
Turnips	4	6	8	9	10
FRESH FRUITS					
Apples	*6*	*9*	*12*	*14*	*18*
Avocados	*4*	*6*	*8*	*10*	*12*
Bananas	*6*	*8*	*10*	*12*	*14*
Blueberries	*8*	*10*	*12*	*16*	*18*
Cantaloupes	*8*	*10*	*12*	*14*	*16*
Cherries	*6*	*8*	*10*	*12*	*16*
Grapefruits	*6*	*8*	*10*	*14*	*18*
Grapes	*8*	*11*	*14*	*18*	*20*
Kumquats	*4*	*6*	*8*	*10*	*12*
Lemons	*4*	*6*	*8*	*10*	*12*
Limes	*4*	*6*	*8*	*10*	*12*

	Very low	Low	Average	Good	Exceptional
Mangoes	*4*	*6*	*8*	*12*	*14*
Melons Honeydew	*6*	*8*	*10*	*12*	*14*
Melons White-fleshed	*6*	*8*	*10*	*12*	*14*
Oranges	*6*	*10*	*12*	*16*	*20*
Papayas	*6*	*8*	*12*	*16*	*20*
Peaches	*6*	*8*	*12*	*14*	*18*
Pineapples	*12*	*14*	*16*	*18*	*20*
Raspberries	*6*	*8*	*10*	*12*	*14*
Strawberries	*6*	*8*	*10*	*12*	*14*
Tomatoes	*4*	*6*	*8*	*10*	*12*
Watermelons	*8*	*10*	*12*	*14*	*16*

2.3.6 A new interpretation of infra-dose sugar efficacy

The current applied research on the insect-repellent effect of low doses of sugar is another example of the lack of understanding of the deep mechanisms of plant-nutrition-parasite interactions. This practice of "sugar protection" has demonstrated its effectiveness in many cases in statistically significant ways, particularly in official trials. Faced with the unexpected success of these experiments, researchers have been led to formulate a hypothesis of antixenosis. While trophobiosis emphasizes the trophic interaction between the parasite and its target, antixenosis emphasizes non-preference related to external characteristics independent of the predator's diet: chemical, morphological, physiological, or phenological signatures, for example. The role of elicitors that low doses of sugar would have on plants is also frequently mentioned. In the experiments conducted, there is little or no mention of trophic preference or repulsion (feeding).

- **Preferring the original mechanism to the copy**

The research program "Sweet" dedicated to this subject did not, moreover, allow understanding the mechanisms of this "sugary" antixenosis (SWEET-Sugars in Field Crops - JB.Thibord, Arvalis 2019). Works focusing this time on the analysis of plant infrared signatures could open up other avenues of explanation. As we have already mentioned, the American entomologist researcher from the USDA (U.S. Department of Agriculture) Philippe S. Callahan (Tuning into nature - 2001) has indeed demonstrated a link between the infrared radiation of plants and insect behavior.

The application of low doses of sugar is interesting in itself because it produces results, but it seems to us that not associating it with its spectral manifestation (NIR) prevents us from discovering the heart of the matter. This technique certainly offers a cost-effective solution to protect crops. However, it is a mask, the semblance of a mechanism that the plant has the ability to activate: indeed, ideally nourished plants produce sap rich in sugars and offer insect predators sensors a typical infrared signature of plants unsuitable for their diet. With the application of sugar, the projected infrared signature is identical, but it only takes rain for the mask to fall and for insects not to be fooled anymore. Instead, it would be better here again to act on the root causes, on the original mechanism rather than the copy. This involves activating efficient proteosynthesis and photosynthesis, which give the plant the profile of a plant unattractive to predators.

2.3.7 Multiple levers for "nutriprotection"

A wide range of parasites operate with their target based on the principle of trophobiosis, which itself depends on the plant's nutrition from the soil. In the case of combating Esca in vines, we emphasize the effectiveness associated with improving root depths. However, root depth is not the only available lever: an active rhizospheric microbiology, whether natural or introduced (*e.g. Bacillus mucilaginosus*) can play a role in transforming molecules in the soil solution, sorting, and acting as a specific transporter.

Our experience has shown us that by acting on the restoration of soil functions, we can improve most of the difficulties encountered by cultivated ecosystems with simple solutions to implement, either alone or in combination.

2.3.8 Plants sickened by our practices

Often, our practices lead to improving the palatability of our crops for parasites. Overall, our practices systematically downgrade trophic chains. In doing so, plant nutrition is also impoverished. This downgrading of soil ecology and foliar applications of pesticides disrupt plant physiology, hinder proteosynthesis, and consequently make plants more attractive and nutritious for various pests and pathogens. Synthetic nitrogen and the use of synthetic molecules, especially chlorinated compounds, tend to reduce protein synthesis and favor proteolysis. This increases the sensitivity of crops.

Photoperiodism, grafting, soil oxygenation level, water supply, mineral balance in the soil solution, differential activation of hormones... It is established that other factors play a role in the major balances between proteosynthesis and proteolysis and should be taken into account. Some phenological stages are associated with dominant proteolytic reactions and thus constitute sensitive phases. This is the case during flowering and early fruiting stages. Plant synthesis functions then partly give way to lytic reactions, including proteins, to supply soluble substances to reproductive organs. Flowering stages often mark the first periods of susceptibility to pests for both perennial and annual plants. It is essential during these periods that the soil be able to provide all the major elements and trace elements necessary for neoproteosynthesis processes.

Phytosanitary practices, especially fungicides, have also significantly contributed to destroying or inhibiting humus-building fungi and other mycorrhizal fungi. Other organisms more adapted to degraded or retrograded soil conditions, and more extremophilic, have taken their place. Behind this shift from the predominance of certain microbial communities to others that are more adapted and aggressive, there are functions that are less well ensured, including those that support the soil-plant food chain. Before the plants themselves, it is first and more directly the soil functionalities that are affected by some of our practices through degradation of fertility levels.

"Farmer-Researchers"

Today, a handful of pioneering farmers sense that something has gone awry in agricultural practices. They embark alone or in groups on experimenting with another form of agronomy on their farms, one that is more "holistic," within highly diversified structures: Holistic Management-Allan Savory, PADV, Earthworm Foundation, Australian Agricultural Sustainability Framework, WAFarmers, Asian Farmers Association for Rural Development, International Association for Agricultural Sustainability (IAAS), etc. Nowadays, it is mainly farmers themselves who are capable of undertaking this work, drawing on the cutting-edge research of scientific organizations, which are structurally and culturally ill-equipped to study complex systems. The writing of this book is also dedicated to these new agronomy pioneers.

2.4 The competitive edge over weeds

Implementation of seedbed preparation, optimal sowing depth, localized fertilization, herbicide treatments, false seedbeds, etc. A significant portion of the farmer's actions aim to provide a competitive advantage to the cultivated plant. However, once soil conditions are no longer balanced in terms of nutrition, structure, water availability, etc., the advantage provided by the farmer diminishes. Factors such as waterlogging, drought, negative redox potential, mineral imbalances, among others, can affect soil conditions. Other plants, including weeds present in the soil, may express their own competitive and adaptive advantages when soil conditions are imbalanced or extreme. The cultivated plant then becomes less competitive for accessing resources. Thus, by restoring balanced conditions in the soil, we have a considerable lever to improve the competitiveness of the cultivated plant in its environment. These balanced conditions also offer potential solutions to the emergence of invasive weeds, or even herbicide-resistant ones. In field trials we conducted, we observed how soil biology restoration practices through nutriprotection measures (see section 2.3.5) reduced the nuisance of infestations of ryegrass and resistant wild oats, while the control remained infested. These soil revitalization practices enhance the competitiveness of cultivated plants.

2.4.1 Limiting plant expenditure through redox management

Recent work, particularly stemming from the research of Olivier Husson (Cirad), suggests a new reason why cultivated plants are more competitive in environments with slightly acidic to neutral pH levels. During their development, plants tend to alter the immediate environment around their roots. In doing so, they promote the maintenance of a slightly acidic rhizospheric pH and, as we will later discuss, a neutral to slightly reduced redox potential. This process of modifying the nearby root environment creates controlled and favorable conditions for plant nutrition and a specific microbial population. In environments where pH (or redox) is imbalanced, plants expend more energy in controlling the root environment. Cultivated plants generally have not been selected for their ability to manage their root environment. Conversely, other species of wild weeds have acquired competitive advantages over millions of years of natural selection to thrive in these types of soils. As a result, in soils with extreme pH levels and inadequate cation ratios, cultivated plants allocate more energy to control the rhizospheric environment. This means that some of their energy is diverted from dry matter production, hence yield, and they are also less competitive against better-equipped weeds. Moreover, this lack of competitiveness can be interpreted by regulatory organisms, which may then seek to destroy them. Maintaining balanced physicochemical parameters in the soil thus contributes to making these crops more efficient and resilient. Mineral soil rebalancing, as we will discuss later, is essential to make fertilization not only a tool for crop productivity but also a lever to access agriculture that is more resource-efficient: reducing herbicides, fungicides, and insecticides. The biggest margin for improving productivity lies in the soil! It is through the soil that this improvement is achieved, not through foliar inputs. Foliar inputs can only come as a secondary measure.

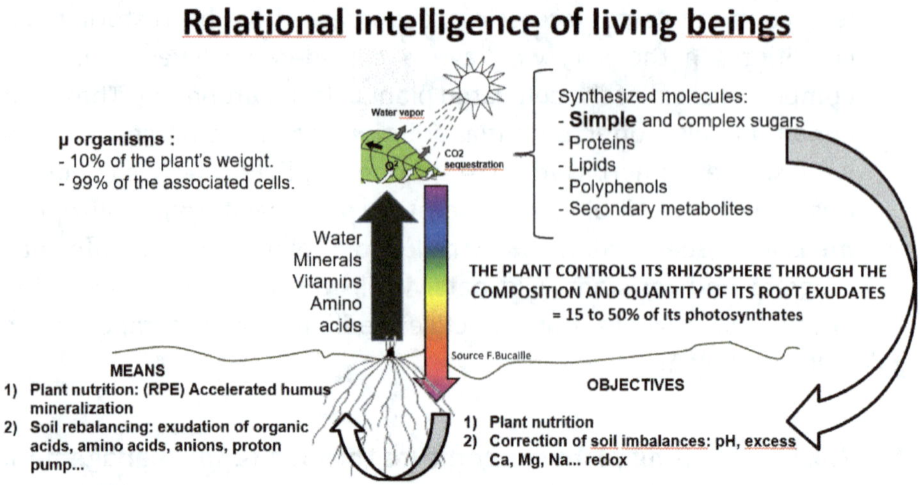

Figure 2.1 – The intelligence of living organisms lies in the fact that plants "invest" a portion of their photosynthetic production through their rhizosphere to ensure: 1) their growth by activating the rhizosphere priming effect (RPE), which leads to humus mineralization and mineral acquisition; 2) the improvement of an unfavorable root environment: pH, redox, mineral imbalances. This function represents an essential expenditure but is not directly productive.

2.4.2 Re-balancing the soils

The advantage that an imbalanced soil provides to weeds is not the only explanation for weed proliferation. Every farmer has likely observed the effect of their practices on soil regarding weed germination. For instance, soil cultivation in wet conditions may favor the germination of chickweed, blackgrass, and knotweeds. However, if soil work is delayed for a day or two to allow for drying, the germination of these species may not occur. Depending on practices, ecological niches suitable or unsuitable for weed germination can be created. Some dormancies are only broken under very specific conditions. In his book *Fascicule des conditions de levée de dormance des plantes bio-indicatrices*, Gérard Ducerf (2015) describes, for various species, sometimes surprising germination mechanisms involving, for example, increases in carbon dioxide concentrations in the soil, as well as increases in methane and negative redox potential. These ecosystem capacities to "awaken" plant species under specific conditions are likely inherited from the system's self-regulatory functions. Weeds are often pioneer plants in imbalanced environments. Their ecological function is to restore more balanced conditions to the environment. In waterlogged soils, for example, dockweed may be competitive in fulfilling this role. With its robust root system, dockweed repairs and restructures soils to improve water circulation. In doing so, it prepares the ground for the subsequent establishment of less hygrophytic plants and a new biocenosis. Through the succession of these transitional biocenoses, the ecosystem gradually balances and naturally progresses towards its climax, the ultimate functional stage. Thus, weeds also have a "role" in the ecology of environments.

Plants are not lazy, but they can be disabled.

> A plant that finds everything it needs in its immediate environment will not cease to grow and explore its surroundings. On the contrary, its tropism is to occupy its ecological niche to the fullest extent. A plant is never "lazy" (though it may exhibit apparent "lazy" signals that need interpretation). Root growth levels are never limited by overly favorable fertility. Therefore, the farmer has every interest in creating the best soil fertility conditions right from the start. Used wisely, so-called "starter" fertilizers can effectively increase plant root levels throughout the entire cycle. However, excessive and unbalanced fertilization can hinder the establishment of a favorable microflora, including mycorrhizal fungi, among others. We will discuss this further in the chapter on fertilization.
>
> There are no "lazy" plants, only plants "deprived of their microbiome."

2.4.3 Pesticides : keeping an emergency kit

We do not irrevocably separate agriculture from the living that links soil fertility and crop protection to the use of pesticide products, whether synthetic or biological (such as essential oils, for example). The last major famine in Europe struck in Ireland in the mid-19th century, causing a million deaths, two million refugees, and two million emigrants (mostly to the United States). The cause? Potato blight (*Phytophtora infestans*) on potato crops, the main food source for the Irish at the time. At that time, there were no effective prophylactic measures. Without waving the threat of such large-scale famines, it remains true that globally, world reserves (carryover stocks) vary annually from only 70 to just over 120 days of consumption. Individually, no country, sector, or farmer can economically afford a year without production. Viruses or other unleashed scourges like rust and blight cannot be fought with herbal teas. Throughout the book, we will see that health is built through balanced fertilization; it can also be maintained with "Low Risk Natural Preparations," but an emergency aid kit must remain available. The widest variety of plant protection products should remain available, from glyphosate to fungicides. These powerful molecules should be managed like medicines, antibiotics, antivirals, or anti-inflammatories for human health: as shields or curative remedies to be used sparingly. It goes without saying that their use should remain exceptional within a justified and parsimonious framework. All medicines are potentially poisonous[7], and their management and use involves

---

[7] Hippocrate : all medicines are poisons (it all depends upon the rate of application)

weighing side effects. The use of pesticides should be approached in the same way, with full knowledge of their mode of action and their effects on cultivated ecosystems and soils. The drift lies in the widespread, systematic, and excessive use of these products, not in their mere existence.

3 • A HOLISTIC APPROACH TO SOIL AND AGRICULTURE

What is the share of truth, of reality, in what is observed? Does what is observed correspond to the reality of things, or rather to epiphenomena, distorted shadows as in Plato's Allegory of the Cave? And if that is the case, could we not trace back to the origin of phenomena by finding the exact causes or approaching them to limit the effect of the distorting mirror? Viewing the world in all its relativity and complexity is a filter that must be imposed on us, especially when it comes to ecology. In this section, we will detail how the soil must be placed at the center of a new holistic approach. This is the only way to initiate a true agricultural "renaissance."

3.1 The birth of holistic agriculture

When tasked with finding solutions to halt desertification in Zambia in the 1950s, Zimbabwean biologist Allan Savory proposed to the government to cull a population of 40,000 elephants that he suspected of "overgrazing" the savanna. A hunting plan was then put in place. The elephants were killed. And what do you think happened? Not only did the desert not recede, but it continued to advance even more rapidly. The facts thus contradicted Allan Savory's overgrazing theory. The biologist could have stopped there or continued to only look at isolated parts of the problem, what we call the Ptolemaic stance. Yet, the biologist acknowledged his initial mistake and endeavored to understand why he was wrong. By shifting to a holistic, "Copernican," and dynamic view of the ecosystem, he demonstrated in the 1960s that grazing and trampling are actually essential for the regeneration of herbaceous vegetation. Otherwise, it suffocates on itself and the rejected patches deprive the young shoots of access to light. Extensive grazing by elephants is not enough to prevent this phenomenon. Only very high-density gregarious grazing (mob grazing) can limit the problem of refusals. This is why the presence of predators is also essential to maintaining the savannah. As soon as the predator disappears, the gregarious instinct dissolves. Bypass zones (rejected patches) appear, preventing young shoots from gaining access to light, more palatable zones are overgrazed until they disappear, and the desert advances. In a stable ecosystem, the presence of each species confined to its own ecological niche guarantees the overall balance that prevails. The system, through its complex interactions, is infinitely richer and more complex than each of its isolated parts.

These observations, corroborated by the earlier work of French agronomist André Voisin (1903-1964), led to the development of the holistic grazing management principle. It advocates for the implementation of rotational grazing with short residence times: 1 day = 1 parcel with high animal densities (1,000 goats or 100 cows/ha). We have experimented with this successfully in Nièvre with our dairy goat herd. In other contexts (mountainous regions), it is the shepherds with their dogs who ensure the maintenance of the animals' gregarious behavior. Grazed ecosystems function better with intensive "flash grazing" management. **When a resource is depleted, it is not always due to excessive pressure on the resource.** When conducted in this manner, livestock farming becomes a decidedly ecological act adapted to steppe, savannah, and grassland areas.

3.2 The need for a Copernican revolution

3.2.1 The Ptolemaic vision: yield as the only reference point

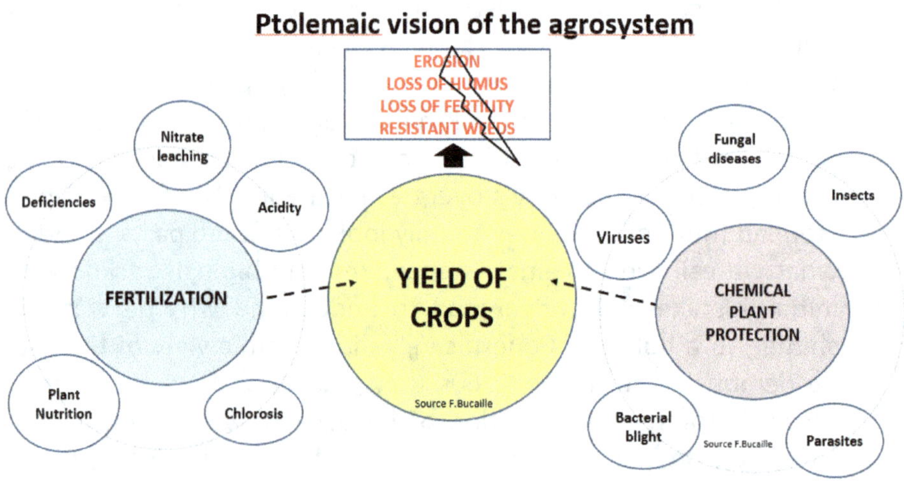

Figure 3.1 – Ptolemaic vision of agriculture : yield at the center of attention.

Ptolemy, Greek astronomer from the 2nd century AD, believed that the universe was geocentric (with the Earth at the center) and immobile. This view shaped much of humanity's understanding of the world for centuries. It wasn't until the Renaissance that Western societies gradually began to accept the idea of a different configuration of the universe. We believe that agricultural practices that prioritize yield above all else, as the sole indicator of success and performance, represent a Ptolemaic vision of agriculture. Yield is merely the consequence of the biochemical and physiological mechanisms of plants. This indicator does not provide insights into the overall mechanisms of the soil-plant system itself. By not considering the microbiological functioning of soils, agronomy has focused solely on observing the effects rather than the causes, shadows rather than the reality of biological phenomena.

As a result, agroecosystems may indeed often succeed in terms of yield, but they suffer in terms of mineral, structural, biological, and functional aspects. This yield-centered vision has long prevented us from linking fertility and plant health. These separate domains of application, focused solely on yield, fail to comprehensively address all components of the soil-plant system: diseases, resistant weeds, soil structure degradation, loss of organic matter, leaching, and so on. Segmented agricultural practices lead to apparent side effects, often positive in the short term (through decapitalization) but mostly negative in the long term. These side effects tend to amplify over time and result in a depletion of our agricultural systems.

■ Excessive focus on above-ground parts

The Ptolemaic reasoning has led agronomy to globally focus only on the above-ground parts of the plant. This is where the value-creating process occurs: photosynthesis. Therefore, this is where humans have identified the economic potential of the crop, especially since grains, forages, and fruits are most often harvested on the surface. Thus, all actions are often directed towards the above-ground part of the plant to maximize yield. Only root vegetable producers, such as beet growers, are sensitive, or at least directly exposed, to soil structure and diseases transmitted directly to the saleable fraction (such as potatoes). The harvesting of these crops also necessitates acknowledging the physical fertility of the soil: drying time, lifting ;kqa&<w[22]speed and soil tare level. Thus, except for this particular confrontation with degraded soil issues, we regularly overlook the heart of the matter: the soil.

■ Limited binary reasoning: one problem, one solution

The issue of combating weeds, diseases, and pests often falls into the equation of one problem equals one solution. Each weed has its herbicide. Each insect has its insecticide or biocontrol agent. Each fungus has its fungicide or biocontrol agent. Often, the solution involves acting by destroying the problem without investigating the causes, without seeing the warning signs that the ecosystem presents. This approach may work for a while, but the ecological cost is significant, and this type of reasoning is not very compatible with agroecology, even if it uses solutions prefixed with "bio". The use of a biocontrol agent is not revolutionary compared to a pesticide: the premise remains the same, namely that pests are inevitable and must be killed.

3.2.2 The Copernican vision: the living soil at the center of the system

- **An agricultural Copernican revolution**

Copernicus, the 15th-century Polish astronomer, developed the theory of heliocentrism (the Sun at the center of the universe). There is always a Sun, an Earth, and all the planets floating in space in a well-regulated ballet. Further afield, there are always the constellations. Simply put, the center of gravity of the World has changed. Where Ptolemy placed fertilizers and pesticides at the center of the system, in the Copernican view, the living soil is at the center!

Copernican vision of the agrosystem

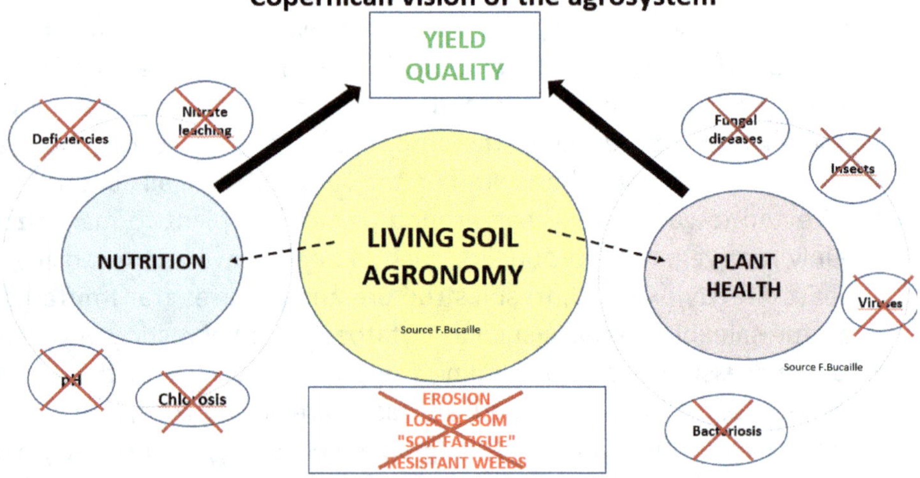

Figure 3.2 – Another vision of the core of sustainable agriculture

> Copernicus demonstrated that the universe was not centered around Earth; rather, it revolves around the Sun in a much larger system. This shift in thinking contributed to the emergence of the historical period known as the Renaissance. It may seem audacious, but with scientific evidence to support it, we are convinced that agriculture is on the brink of a Copernican revolution. The center of the system must also change. By considering what happens in the soil—the living—then we give ourselves the opportunity to strengthen plant health and eliminate or significantly limit the impact of a large portion of pathogens, all while increasing yields.

■ A holistic ambition

"In ecology, a holistic approach allows for a better understanding of the interactions between living beings and the rest of the ecosystem of which they are a part. An element or entity (molecule, organelle, hormone, organ, organism, superorganism, population, ecosystem, biome, etc.) is understood based on its position, relationships, and activity within the organism, landscape, or biosphere. Two principles illustrate ecological holism: the change of an element will necessarily affect, in the short or long term, the entire system, and the change of a system will necessarily affect, in the short or long term, all the elements constituting it."[8]

The approach to agriculture that we advocate for is to consider crops as unstable ecosystems supported by soil, plants, and climate. This holistic approach leads us to rediscover connections that our modern practices have overlooked. About 80% of these connections are established in the soil, which is the new center of gravity.

3.3 Removing obstacles

The Ptolemaic vision of agriculture did not prevent our societies from achieving the first agricultural revolution. It is largely capable of providing sufficient calories and protein to feed the entire human population. However, this revolution was based on principles that paid little attention to the living soil system. There are numerous unintended side effects. Therefore, putting soil at the center of the system represents a genuine mental revolution that it is desirable to implement today. We inherit an entire organization of research, advisory services, and thought patterns. This inheritance is an opportunity for the development of a new agriculture, as extremely specialized research has been conducted. These rich findings deserve to be brought out of the "Plato's cave" and into the light of other knowledge. This rich agronomic heritage, to fully express itself, must go beyond the deeply ingrained mental maps that have always separated laboratories, cooperatives, trading companies, and institutes into two departments: "Fertilization" and "Plant Protection." These departments have organizational charts, mental structures, and real estate architectures based on this distinction. However, the links between nutrition and health are evident and well-documented.

---

[8] "Examples of theoretical applications" (translated from French), in Wikiwand, *Holisme*, [online], www.wikiwand.com/fr/Holisme.

3.3.1 Finding new operating methods

Change in the agricultural model will not happen by opposing the old one, but by building on the knowledge and expertise acquired. The task ahead is to shift paradigms, but certainly not to discard everything. It's about rediscovering the overlooked connections, especially those with the soil, as 80% of productive and healthy agriculture is constructed there.

The agronomic shift and the focus on soils must be accompanied by an evolution in research methodologies. Current and past methodologies have generally deprived and continue to deprive global agriculture of interesting results based on the natural functionalities of soils and plants. It is quite urgent for research institutes to become more agile in adapting protocols to these objects of study. Even laboratories with the most advanced approaches to considering the ecology of the soil-plant system still underestimate many parameters related to living organisms, soil, and plant nutrition quality.

Above all, the soil must be able to perform its own essential functions and provide balanced nutrition for plants. Experimental models that do not link soil health, soil life, and fertility are thus poorly suited to produce significant results concerning ecobiology. New methodologies need to be developed. Rarely do we invest in controlling soil factors, the spatial mobility of living organisms, or the transportation of minerals, molecules, and microorganisms by fauna, yet this is such a central issue.

3.3.2 Measuring the "living soil" effect

The farming practices we have inherited largely stem from implementation tests conducted in micro-plots (3×7 m). These micro-plots serve as the "lenses" through which applied research has observed and continues to observe crop behavior. Plant protection products, fertilization models, new varieties, and many other practical applications in agriculture derive from these statistical selection processes in micro-plots. By adopting this single filter, we have been able to develop certain techniques, but we have also deprived ourselves of hundreds of others and all those that involve soil life. Indeed, the use of statistical methods such as randomization and other Latin square designs often cannot compensate for the limitations of micro-plots in highlighting the complex and diffusive factors related to living organisms.

- **Micro-plots are well calibrated for "cides"**

Let's take the example of comparative trials of fungicides and natural defense elicitors (NDE) in micro-plots. The disease-inoculated controls, which do not receive treatment, are in direct proximity to other micro-plots. However, these controls can potentially serve as sources of infestation. While a synthetic product provides total and effective protection near the infested control, natural defense elicitors (NDE) may not offer the same level of protection. However, this doesn't necessarily mean that they lack efficacy under "normal" field conditions. Simply put, their effectiveness in real-world settings cannot be evaluated on such a small scale. Research conducted in micro-plots thus unintentionally selects for the most "aggressive" solutions and those least connected to the soil's ecobiological phenomena, whose boundaries extend beyond those of micro-plots. For example, experiments by Suzanne W. Simard have highlighted the extraordinary mobility of phosphorus in forest ecosystems on a timescale shorter than that of cultivated plants. The activity scale of an NDE is much larger than the spatial scale of a micro-plot due to the exchange mechanisms it integrates into the soil. We believe that trials involving NDE, as well as soil-focused cultivation practices and soil activators in annual crops, should be conducted on plots with a minimum width of 12 meters and preferably 24 meters. When a soil prebiotic that rebalances a microbial ecosystem, such as an azotobacter-type inoculant, is tested, the risk of transfer between modalities is significant. The risk of reaching the conclusion that there is no significant difference is common, as the control itself will have also benefited from cross-inoculation. Current experimental setups are designed to destroy forms of life deemed undesirable (insecticides, fungicides, etc.). They have not been calibrated to evaluate revitalization products or practices. We know how to "devitalize" a specific area, but we have much more difficulty preventing the circulation of a living organism or circumscribing the solidarity mechanisms of life in the soil. Whether it concerns a beneficial or pathogenic microorganism, effective confinement is one of the most challenging exercises in experimentation. The figure below shows the positive effects on vegetation (observed by satellite imaging like Farmstar near Angers) of the application of azotobacter compared to its control (without azotobacter). We have scaled a classic experimental plan of micro-plots in a Latin square on this figure. We can see that the "migration" of bacteria is too significant to allow us to see the reality of its action, which is nevertheless striking on a larger scale. We condemn ourselves to non-significant differences (NSD) in conclusion.

Observation of an experimental design using a Latin square. 5 treatments with 4 replications. Dimensions of each micro-plot: 3 m x 10 m

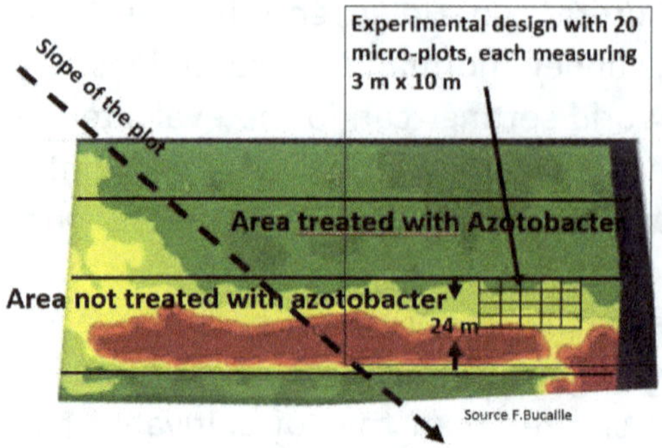

This aerial photograph, taken in infrared and then processed with software (FARMSTAR), reflects the actual, measurable effects of Azotobacter chroococcum a few months after application.

The sinuous and random appearance of the reflectance curves illustrates how the boundary of the beneficial effects of the inoculation deviates from the straight edges of the application limits, which were initially strictly outside the rectangle representing a 24-meter untreated strip.

Figure 3.3 – Effect of azotobacter application on a Latin square experimental plan (5 modalities in 4 repetitions) on micro-plots. Plot near Angers: experimental plan with micro-plots of 3 × 10 m

> This aerial photograph captured in infrared and subsequently processed by software (FARMSTAR) reflects the real, measurable, and measured action of *Azotobacter chroococcum* a few months after application. The sinuous and random appearance of the reflectance curves shows how much the limit of the beneficial effects of inoculation moves away from the straightness of the spreading boundaries that were initially strictly outside the rectangle representing a 24 m untreated strip.

▪ Taking the "Internet of the soil" into account

In and on soil, there exists a multitude of "carriers," such as nematodes, ground beetles, earthworms[9], fungal mycelia, the root system itself, water tables, and the splash effect of droplets (Yang, 2017). For example, it has been demonstrated that *Pseudomonas aeruginosa*, a flagellated bacterium, can move up to 20 mm in 24 hours in moist soil, by its own means (cf. Griffin & Quail, 1968). With this in mind, is it incongruous to ask whether a bacterial colony introduced into soil could move 5 m in the space of six months, generation after generation? The more sandy and porous the soil, the more effective the mobility of bacteria (due to fewer attractive forces of clays and more lateral circulation channels). Additionally, a study from MIT (Massachusetts Institute of Technology) published in *Nature* showed the significant role of rain in the dissemination of soil bacteria (Joung, 2015). After laboratory simulation of various forms of rain on different soils (sand, clay, etc.) and high-

[9] cf. Daane, 1996 ; Kohlmeier *et al.*, 2005 ; Warmink & van Elsas, 2009 ; Glaeser *et al.*, 2016.

resolution camera imaging, the images show that after impact, each water droplet rebounds as aerosols. These aerosols have trapped several thousand bacteria from the soil and can drift for a kilometer. Other studies have demonstrated (cf. Kellogg & Griffin, 2006) that wind can carry bacteria over even greater distances, from 1 to 25 km, carrying solid particles with a high chance of survival, especially for bacteria that can sporulate (e.g., *Bacillus mucilaginosus, amyloliquefaciens*). Based on MIT simulations, it turns out that 1-25% of soil bacteria could be dispersed in this way, or, according to their calculations, some 10,000 to 800,000 billion billion bacteria per year. What an amalgamation ! For physical reasons, this phenomenon is even more pronounced on clay soils than on sandy soils, and on all types of soil at positive air temperatures (the phenomenon is greatest above 25°C).

Drawing inspiration from applied research on pheromones

> Applied research has shown that it is sometimes capable of greater agility in adapting its protocols to its subject of study. Otherwise, current methods of control known as mating disruption would never have been developed. Indeed, the pheromones used to prevent encounters between males and females of various pest species are detectable by targets over several tens of hectares. Anyone can easily imagine what the evaluation of these pheromones would have yielded as results on the scale of micro-plots. In this field, as it concerns airborne phenomena, the evidence of the phenomenon is less counterintuitive than for the soil.

Using precision agriculture judiciously

> We have already seen the extent to which micro-plot trials are ill-suited to highlighting the impact of practices based on the soil ecosystem. With the emergence of onboard sensors aimed at mapping soils, we now have the means to better account for the bias introduced by soil heterogeneity. We can now position trials not in micro-plots, but with modalities extended to half, third, or quarter plots. Samples are then taken, harvested and weighed in areas "representative" of soil heterogeneity, based on maps of electrical conductivity, humus content and acidity. It is thus possible to compare different modalities in areas with equivalent physicochemical characteristics, with the added benefit of being able to evaluate the performance of a product in areas of different potentials, while also eliminating the risk of proximity interference. Very promising work is underway within companies like Geocarta or Precifield, which will improve experimentation processes. Collaboration with statisticians is needed to calibrate experimental protocols with probably fewer repetitions but on larger plots. Inter-plot meta-analyses already provide solid scientific contributions to observations in real-life situations on farms to compare different practices.

3.3.3 Taking into account the relativity of phenomena

■ A multifactorial approach is necessary

Applied research struggles to break free from the habit of working solely on the criterion of yield in "controlled" situations and to apprehend ecological and physiological phenomena through an accounting and compartmentalized approach. In ecology, the number of variables at play is enormous, and everything is inevitably subject to interpretation and multifactorial.

Microorganisms like azotobacter are capable of fixing atmospheric nitrogen (N_2) directly from the soil atmosphere. However, this free-living microorganism has certain requirements. It is strictly aerobic, meaning that a soil without porosity will not suit it. To perform its nitrogen-fixing functions, its tool, the enzyme nitrogenase, requires activation by molybdenum and cobalt. Additionally, to build its own cells, DNA, and organelle components, it requires sulfur. Finally, a soil saturated in nitrogen will tend to lead the plant, which is supposed to host and feed our bacterium through its exudates, not to provide it with nourishment. Plant-specific regulatory mechanisms will dictate not to favor a biological mechanism that would lead to overconcentration in an element it already abundantly possesses. In the background, a crop lacking certain nutrients at a perceived limiting threshold will not fuel a bacterium supposed to provide nitrogen for superior growth with carbohydrates. If, for example, phosphorus is limiting, the plant will not invest in this auxiliary aid such as azotobacter, as the bottleneck on the phosphorus element condemns the expenditure (towards azotobacter) to a non-return on investment. Few experimental platforms can manage all these contingencies correctly. This is perceived as creating biases, and introducing too many factors that are difficult to analyze when it comes to identifying what really worked. Yet the answer is simple: it is the whole that works.

■ Promises of metagenomics

Research should, in this sense, adopt a greater relativity towards living organisms in the interpretation of its experimental results and shift its practices to better account for it. This is, for example, the essence of research conducted in the field of metagenomics. This approach, enriched by enzymology and nutrigenomics, allows for a better representation of the profile and balance of soil functionalities. This new science, based on DNA analysis and metabarcoding, focuses on the study of the overall functioning of complete populations of microorganisms. It starts from the observation that many microbiomes form an interconnected whole from which it is not possible to isolate or export a part without seeing it decline. This is because the mechanisms of cooperation and exchanges are so intense and vital for each of the organisms composing the biota. We only know about 10% of soil microbiology. The reason is simple: we have only worked on life forms that support pure in vitro culture, separately, on an artificial culture medium. However, the vast majority of microorganisms require, from a nutritional point of view, polyketides or obligatory exudates from other organisms: plants, bacteria, fungi. Conditions that we cannot reproduce in controlled environments.

■ With archaea, we discover a major aspect of the nitrogen cycle

The sole difficulty of observing certain microbial species in pure cultures has led to the neglect of certain families. For some, it was simply assumed as obvious that they were unlikely to be found in cultivated soil! This is the case for archaea, which were previously supposed to be present only in extreme environments (deep sea trenches, geysers, etc.). After searching for them, it turns out that they are present everywhere: in aquatic environments, the rumen of cows, and soils (Leininger et al., 2006). Sought through specific markers of *Crenarcheotae* (GDGTs: glycerol dialkyl glycerol Tetraether and crenarcheol) and correlated with an ammonia oxidation enzyme (amoA), it turns out that these archaea are present in all soils, even at great depths. However, archaea are very active in the nitrogen cycle. One of the main phenomena in the nitrogen cycle - that of the oxidation of ammoniacal forms of nitrogen - was until recently considered mainly the work of ammonia-oxidizing bacteria *Nitrosomonas*, thanks to a dedicated enzyme, ammonia monooxygenase (amoB). Today, we discover that the impact of archaea in this oxidation function is probably major and even more significant than that of *Nitrosomonas* bacteria. This is a crucial discovery because a major player in a cycle as biologically and economically important as that of nitrogen (the carbon footprint of nitrogen fertilizers alone represents up to 50% of the total footprint on farms without livestock) had not been taken into account. Yet it has specific requirements, particularly in terms of minerals and trace elements. This is the kind of new information we're incorporating into our thinking and advice.

Soil science is an incredibly rich field for the study of these microbiological communities. Soil microbiology is a largely unexplored universe from which we can rapidly reap the benefits. New reservoirs of productivity are already opening up to us, and this time they are inexhaustible if we know how to nurture and respect them.

3.3.4 Expanding plant models

■ Taking soil connections into account

Many experiments are conducted on plant models, with one of the most commonly used being *Arabidopsis thaliana*, commonly known as thale cress or Arabidopsis. It belongs to the Brassicaceae family. This plant serves as a model organism in biology laboratories. Arabidopsis thaliana is a perfect model organism due to several advantages:

- Small size (20 cm), making it easy to cultivate in growth chambers.
- Rapid multiplication (six weeks from germination to maturity).
- Small genome (rice and corn are four times and twenty times larger respectively).
- This plant can also be easily genetically transformed using an agrobacterium (*Agrobacterium tumefaciens*).
- No economic importance, so observations are easily transferred between researchers in the global scientific community.

The inconvenient truth is that Arabidopsis thaliana belongs to the 5% of cultivated plants that do not form mycorrhizal associations (including rapeseed and beet as well). However, mycorrhization is one of the most beneficial symbioses possible for cultivated plants. Unfortunately, this ability never appears in the knowledge we advance based on this plant model. At the very least, in the coming years, we will need to study another plant model whose connections to the soil express more complexity.

Even for other mycorrhizal species tested, to avoid "bias" in experiments, the soil is often disturbed, mixed, homogenized (which obviously breaks any remaining mycorrhizal mycelium) and sometimes even sterilized to ensure controlled conditions. Instead, we should systematically include these synergies and symbioses to select the varieties that host them best and detect fertilizers and practices that promote them or at least do not hinder them. And of course, we should avoid all practices that are unfavorable to them.

■ Varietal research: taking into account the plant's ability to engage in soil solidarity

It should be noted that varietal research has also focused primarily on the criterion of yield. In so doing, it has favored plants capable of transferring all their production as efficiently as possible to the "cash" organ. Plants that contribute to the soil ecosystem through their exudates have often been penalized in selection protocols. Yet plants that participate in and maintain activity around the rhizosphere are also in a position to be in better health, protected by an advanced level of nutrition and better nutritional value for humans. Varietal research also deserves to undergo a Copernican revolution.

3.4 New forms of agriculture have to evolve

Organic farming, conservation agriculture, permaculture, agroforestry, third-way agriculture, integrated agriculture, etc. In recent years, other agronomic currents have emerged, often constructed in opposition to conventional agriculture. The shifts initiated by these movements are sometimes excessive and do not avoid new pitfalls. In any case, these pioneering agricultural approaches cannot escape the principle of reality and agronomic fundamentals.

Organic farming, which rejects the use of synthetic products as a matter of principle, itself sometimes falls into other pitfalls. Massive doses of copper or the recurrent use of essential oils (orange essences to combat mildew, for example) have a highly negative impact, particularly on earthworm populations. Copper solutions have a lasting effect on soil degradation. Copper, a heavy metal, is neither leachable nor degradable (by definition, since it is a simple element), so its half-life is eternity (unless transmutation is considered...). Not all natural products are harmless by nature. The road to hell is paved with good intentions...

Permaculture has the merit of taking ecosystems and soil life into account from a global perspective. However, its implementation often remains confined to very small-scale structures. These experiments are clearly of interest on a human level, in terms of shared values, but they can in no way be a precursor model of what the agriculture of the future will be like to feed the world. It is an illusion to present certain examples based on more or less remunerated labor and a transfer of fertility from forests (RCW i.e. Ramial Chipped Wood and bark or pruning compost), from exhaustible resources (peat) or from organic by-products from conventional agriculture (straw, manure, compost) as a generalizable model.

3.4.1 Conservation or regenerative agriculture

In light of the evidence of the excesses of modern agriculture, it is tempting to do exactly the opposite of what has been practiced. Drawing from this line of thinking, models of agriculture are emerging based on a "moral" principle aimed at intervening as little as possible to mimic the functioning of the natural ecosystem to the fullest extent possible. **Non-tillage or permanent cover are thus key principles underlying this new agriculture.** However, in terms of the environment, the expected increases in biodiversity or organic matter in the soil do not always occur systematically, as shown by the long-term study on the effect of soil tillage on biogeochemical cycles conducted by Inra and Arvalis (Bassem Dimassi, 2014). These practices, when applied without first checking some key success factors (macrofauna, swelling clays, humus, etc.), sometimes lead to degradation, especially when the soil on which they are applied lacks the necessary capacities for self-restructuring. These systems also often lead to a climax lag, which we will examine further, that hinders the expression of full efficiency.

3.4.2 Biocontrol: a step forward or the "green extension" of a defensive vision?

Today, biocontrol represents a new frontier for many companies. According to the definition of the French Academy of Agriculture, it encompasses "all methods of plant protection that use natural mechanisms. It aims to protect plants by favoring the use of mechanisms and interactions that govern relationships between species in the natural environment." Biocontrol mobilizes agents that can be classified into four categories:

- Macro-organisms (ladybugs, trichogramma wasps, thrips, etc.).
- Micro-organisms (*Bacillus subtilis, Bacillus thuringiensis*, etc.).
- Pheromones (sexual attraction of predators).
- Natural substances of animal, plant, or mineral origin (essential oils, copper, etc.).

The last three categories are considered phytopharmaceutical biocontrol products and must obtain marketing authorization and therefore be listed in the "Phytosanitary Index" of protective products. These agents have proven effectiveness but are subject to some weaknesses:

- The sustainability of the proposed solutions: development of resistance or arrival of new aggressors.
- There are still a number of unresolved issues: weed management, simultaneous attacks by several pests (e.g. corn borer and stem borer) or certain "orphan" diseases such as downy mildew.
- All these solutions are of the "cide" or "fuge" type, and do not necessarily

include active soil revitalization.

These solutions have the immense advantage of offering alternatives that are generally less harmful to the ecosystem than "old" synthetic molecules. However, their main risk is also to encourage us to "look at the sage's finger rather than the moon." By focusing primarily on parasitism and diseases rather than the central gravity of health: the soil.

3.5 The risk of making something new out of something old

We are witnessing the emergence of solutions intended to revolutionize agriculture in an ecological direction. One example is biocontrol tools. However, these new tools are often implemented based on the same principles as the old ones. We combat the symptom, suppress it, mask it, inhibit it, or kill it, but we do not listen to the signal sent by the ecosystem. Replacing a synthetic pesticide with a biocide or biocontrol, or even a decoction, presupposes that a plant is naturally destined to be attacked by insects, fungi, and viruses. Confronting parasitism with natural solutions is certainly less harmful to the environment, but it still does not prioritize strengthening the health of the plant. We still do not make the link between fertility and health. It remains a problem/solution reasoning without restoring the vitality of the agroecosystem as a whole. The risk of not implementing the Copernican revolution is to push the old system even further, towards other frontiers, and to not face reality head-on. Then we would content ourselves with making something new out of something old and deprive ourselves of the tremendous biological potential that we have beneath our feet.

4 • MOVING AWAY FROM A SIMPLIFIED VISION OF SOIL FERTILITY

Where does the rule of three units of nitrogen to produce a deciton of wheat come from when it exports less than two? How do we explain that we base soil fertility mainly on three elements, NPK (nitrogen, phosphorus, and potassium), while the plant is composed of many others? Most current agronomic rules have been built on foundations that largely ignored certain processes concerning soil fertility. The palpable consequence of all this is that very often in the field, agronomic theories are contradicted. Predictive prophecies do not materialize, and others, unexpected, arise. In terms of fertility, reality often seems to defy accounting and mathematical laws. In this section, we will present an analysis of various biases that have resulted in oversimplifying questions of fertility and fertilization.

4.1 Not-so-complete fertilizers

Agronomic practice has simplified the question of soil fertility. This simplification is accompanied notably by the implementation of inadequate and misleading semantics in the field of agronomy. Let's take the case of NPK fertilization, nitrogen (N), phosphorus (P), and potassium (K). Fertilizers containing these three elements are known as "complete fertilizers". However, reducing fertilization to these three elements alone is certainly not qualifying as complete fertilization. Plants mobilize much more diverse and complex soil resources that are found in plant tissues. Thus, mineral exports by crops are much larger than commonly assumed. On average, crops export 25 to 30 different elements. However, we only provide the soil with between three and five elements (NPK, calcium, and magnesium most often). The inputs of trace elements are quite systematically presented as optional, which they certainly are not. Moreover, so-called complete fertilizers are becoming less and less complete. Over time, advances in refining techniques, the choice of fertilizer sources mean that NPK inputs are considerably impoverished in trace elements. Dephosphorization slag, tricalcium phosphates, chalks, natural phosphate sands, calcined bone powders, and other refinery blacks have been replaced by superphosphates made from pulverized ores, rid of heavy metals and uranium when present, and enriched with phosphoric acid. Cultivated systems have thus lost the benefit of combined inputs of trace elements, notably bioavailable manganese, which we have already seen above as one of the beneficial functional aspects. Additionally, slag provided, besides phosphorus, numerous trace elements from steel processing, which were interesting for restoring the fertility of degraded soils and correcting the effects of surface leaching and acidification. Pure potassium chloride has been substituted for seaweed ash, peat, and wood ash as a source of potassium. As for nitrogen, it is produced by transforming atmospheric nitrogen into perfectly pure ammonia using the Haber-Bosch process. These inputs of mineral nitrogen have replaced materials with complex compositions such as guano, Chilean nitrates, sewage waters, and ammonia waters condensed and stabilized by sulfur (forming ammonium sulfate) from coal carbonization. Once again, the use of increasingly pure minerals has reduced the combined inputs of elements that were often beneficial for soils and crops.

4.2 Fertilization : a better calibration needed

The balance method still prevailing today to adjust fertilization to plant needs only incompletely considers soil reserves and does not take into account all of a plant's needs, including trace elements. It is based on an accounting principle of balancing income and expenses, in other words, providing as much as has been exported. The layer analyzed is limited to the top 20-25 centimeters of soil. This is very restrictive because most of the time it is much less than the plants' exploration potential. While the top soil layers contribute more intensively to the feeding of the cultivated plant than other horizons, it is not rigorous to ignore the remaining potential. Wheat roots, in some favorable situations, are capable of exploring up to 2.5 meters of soil column! In these situations, how could a soil analysis conducted in the top 25 centimeters be the sole decision tool for fertilization?

4.2.1 Surface rooting

The corollary of this too superficial and not sufficiently volumetric understanding of soils is that, through habit and a lack of attention to indicators other than the chemical richness of soils in terms of major elements, we have allowed practices to develop that do not encourage deep rooting. Luxury doses of phosphate inhibit mycorrhizae, which are supposed to extend plant roots, extracting and transporting minerals that are inaccessible due to their distance from plant roots or chemical unavailability. In this way, we have progressively developed a form of agriculture that is less and less autonomous, and less able to express the effect of terroirs in quality chains. For foods (wines) produced in areas with controlled designation of origin, we believe it is vital to rediscover this 3D vision of soils and restore correct rooting depths. This is a differentiating advantage, without which agricultural production would be doomed to be integrated into volume chains, with supply trivialized and valued at world prices.

4.2.2 Fertility reservoirs

The volume of soil explored is not the only fertility reservoir neglected by the balance method. Multiple biological mechanisms are capable of providing significant chemical fertility to the soil. The activity of free-living diazotrophic bacteria (such as Azotobacter) or rhizobia, which can fix atmospheric nitrogen at rates of 30 to 40 kg of nitrogen/ha/year (equivalent to 100 kg of ammonitrate), is almost never taken into consideration. Similarly, the activity of certain bacteria capable of providing phosphorus, potassium, and silica, available to plants from chemically insoluble forms, is often overlooked. As the activity of these auxiliary microorganisms is considered negligible, the indirect effects on them of substances and molecules used in agriculture have not been evaluated. Taking only the example of nitrogen-fixing bacteria, they are sensitive to excess synthetic nitrogen, which completely inhibits them. Massive inputs will therefore make plants 100% dependent on external nitrogen inputs. Another example: copper inputs, in the form of Bordeaux mixture for fungicidal purposes, are particularly harmful, even destructive to diazotrophs beyond values exceeding 50 ppm in soils (we find soils exceeding 200 ppm with records at 800-900 ppm). This effect will last for decades, given the immobility of the copper ion. These are just a few examples of unintended effects that harm the economy of a farm, whether organic or conventional.

Over decades of experimentation, models and standards of recommendation have been based on dysfunctional soils, on crops with less efficient root systems, on inhibited and/or unbalanced microbiologies. What was abnormal has become increasingly common. The common has become the norm. This new, man-made norm has given rise to a model that drives both advice and experimental systems.

4.2.3 Forfaiting losses

Finally, the balance sheet method is based on accounting "lump sums", which systematically include an element of "unknown markdown" for moving items such as nitrogen. In store accounting, shrinkage is the term used to describe products which disappear from stock without explanation, and which in reality are more often than not stolen. Well, the calculation of requirements systematically includes around 70% markdown for nitrogen (see box). No store would accept this. Why should the cultivated ecosystem systematically be so inefficient with regard to nitrogen? Aren't there ways of understanding these losses and reducing, if not canceling them out? We will provide answers to these questions, and solutions exist to reduce these losses and make the cultivated ecosystem a winner with regard to these phenomena. Calcium/Magnesium ratios are one cause, active microbiology another, and rooting depth and density a third.

Table 4.1 – Standard nitrogen requirements estimation in units (kg/ha of pure element) for a deciton of exported crop (source: Arvalis)

	Yield objective (coefficient b)	**Quality objective (coefficient bq)**
Winter wheat (excluding hard wheats)	2.8 to 3.2	3 to 3.2
Hard wheat	–	3.5 to 4.1
Durum wheat	–	3.5 to 4.1
Winter barley	2.5	2.2
Spring barley	2.5	2 to 2.5
Triticale	2.6	–
Oats	2.2	–
Rye	2.3	–

A flat-rate loss of 70% in the soil's black box

It is commonly accepted that to balance the nitrogen fertilization equation for wheat, the nitrogen N requirement per deciton of wheat produced is 2.8 to 3.2 kg/q (recommendation from the Union nationale des industries de la fertilisation (the National Union of Fertilizer Industries) [UNIFA] and Arvalis). However, for wheat with 11% protein content (average case in France), considering that the mass fraction of nitrogen *N* contained in plant proteins is 16%, then one deciton of wheat corresponds to the export of: 11 × 0.16 = 1.76 kg of nitrogen *N*/deciton.

There is a 70% gap between estimated needs and nitrogen exports (in the case of a recommendation at 3 units/q: (100 – (3/1.76) × 100 = -70%).

The doses used and the nitrogen use efficiencies also prove disappointing on all continents and for all crops (Krupnik et al., 2009).

In wheat, 31% efficiency, corn 24%, and rice 37% (percentage of nitrogen applied found in the plant) due to losses through leaching/volatilization and transfers between soil pools.

4.2.4 Secondary functions of fertilizers often overlooked

During World War II, the Allied army injected significant quantities of ammonium salts into fields. What was their purpose? Certainly not to fertilize the soil. The army actually used ammonium for its compacting effect on soils. It was about stabilizing, at a low cost if you will, the locations of roads, field camps, or landing strips. Similarly, in large civil engineering projects, the same technique is used with ammonium salts (Davidson, 1962). The expected functionality in agriculture from ammonium inputs is quite different. It is mainly about ensuring the nutrition of the cultivated plant. But can we neglect the secondary effects induced for the cultivated system? Every action on soils, whether mechanical, chemical, biological, etc., is usually reasoned around a specific, precise, and often unique functionality. But each targeted action carries with it disruptive secondary functionalities that are sometimes beneficial, sometimes negative, and often multisite. Everyone will easily understand that intense, repeated use, beyond the strict needs of crops, will produce the same effect sought by civil engineering engineers: stabilization, that is, soil hardening, compaction, and the need for more power to work them. Still concerning nitrogen fertilization, synthetic nitrogen inputs in the form of NO_3 also have an effect on plant hormonal balances. Nitrogen favors cytokinins and disfavors the hormone that is antagonistic to it, auxins. Elongation phenomena are favored, while rhizogenesis is conversely disfavored. Excessive use will contribute to reducing rooting depths and the ability of crops to explore deep horizons and benefit from the soil's water and nutrient reserves.

- **Mineral balances are disrupted**

Synthetic nitrogen inputs also alter the absorption of other minerals. They promote the plant's absorption of nitrogen, as well as calcium and magnesium. However, they penalize the absorption of phosphorus, potassium, copper, zinc, manganese, and boron. This does not mean that agriculture should do without these synthetic fertilizers. But the collateral effects must be known and better taken into account. Early-season nitrate inputs - in spring - provide abundant nitrogen nutrition while the soils are still cold. Copper, zinc, manganese, and molybdenum will not yet be very available due to barely awakened biological activity. The sap composition is therefore likely to be imbalanced, especially since their absorption will be hindered by the antagonism exerted by abundant nitrates. These trace elements are involved in numerous enzymatic reactions, so the plant is unlikely to be able to ensure efficient protein synthesis. Nitrates will remain in this form or in the form of simple amino acids for longer and will sensitize the plants to fungal diseases. **Poorly managed nitrogen fertilization prepares the crop to "value" the use of fungicides by making them indispensable.**

▪ Ammonium fertilizer can decalcify and acidify

A look at laboratory analysis techniques is also enlightening. Indeed, the so-called Metson method for calculating cation exchange capacity (CEC) involves using a reagent based on ammonium acetate. The NH_4 (ammonium) it contains is capable of substituting for other cations held by electrostatic attraction on clays and humus, such as calcium, magnesium, potassium, and sodium. The demonstration by this laboratory technique used worldwide on all types of soil shows how powerful and universal this reagent is in its function of substituting Ca, Mg, K, and Na cations. The use of ammonium fertilizers in quantity can therefore release stored cations into the soil solution. They are reputedly non-leachable, for a reason; they take the place of others on the CEC. Initially, the consequences seem eminently positive since an excess of minerals in the soil solution can result in yield increases. These are real, but result from decapitalization, a "mining" extraction of soil resources. The rest of the scenario is the nitrification of this ammoniacal form in two steps, nitritation and nitration, with the following chemical reactions:

1. Nitritation: $NH_3 + O_2 \rightarrow NO_2^- + 3H^+ + 2e^-$
2. Nitration: $NO_2^- + H_2O \rightarrow NO_3^- + 2H^+ + 2e^-$

The oxidation process of mineral nitrogen (ammonia, ammoniacal nitrogen from ammonium nitrate, or urea) by soil microbes is also itself a producer of hydronium ions (H^+ sometimes also expressed as H_3O^+) responsible for a decrease in pH. After nitrification, ammonium displaces the bases of the clay-humus complex, leaving a trail of hydronium ions behind.

4.2.5 Nitrates never travel alone

Mineral nitrogen fertilization can be implicated in many other secondary effects that are still largely ignored, particularly in terms of losses in calcium and other cations. Indeed, mineral nitrogen inputs lead to the presence of abundant nitrates in the soil. However, nitrates never travel alone. They form with calcium, potassium, or sodium cations salts such as calcium nitrates, etc. These salts are extremely soluble and highly leachable. Thus, nitrate leaching also results in the departure of cationic "bases," amplifying surface acidification. Improving the efficiency of nitrogen utilization by the plant not only saves nitrogen fertilizer but also reduces the need for liming to rebalance the pH! Nitrogen, including synthetic nitrogen, is not inherently harmful to the cultivated ecosystem if this nitrogen is efficiently used by the plant. It is the unused nitrogen that acidifies and "ages" the soils. According to the work of Professor Philip Barak (1997) from the Department of Soil Science at the University of Wisconsin (United States), 40 years of mineral nitrogen fertilization cause aging equivalent to 1,000 years of "natural" aging. This is due to acidification (an average decrease of 1.5 pH points), clay structure degradation, and a 30% reduction in cation exchange capacity (CEC).

■ Side effects associated with the use of potassium chloride

Using the balance method without considering the biochemical mechanisms in soils can lead to another agronomic error when using potassium chloride (KCl). Indeed, chloride forms in the soil with cations such as calcium, calcium chloride salts, which are highly water-soluble and hygroscopic, and therefore extremely leachable. The higher the doses of KCl, the greater the losses of calcium as CaCl through leaching, thus creating an obligation to liming to compensate. This phenomenon is even more serious and pronounced in no-till systems, as there is no longer recycling of bases to the surface through tillage. We often observe significant surface acidification. Potassium sulfate does not have such a widespread effect since the new salt created by the combination of the sulfate ion with the soil calcium, calcium sulfate, is significantly less soluble (2 g/liter) in water than calcium chloride (745 g/liter). Potassium chloride is not only a good decalcifier, as we have just seen, but also a good soil acidifier because potassium has the property of displacing H^+ ions in the soil solution, taking their place on clays and humus. In fact, potassium chloride (or sometimes calcium chloride) is used by soil analysis laboratories to displace H^+ ions from the CEC. That's why pH KCl values given by laboratories are always lower than pH water.

4.3 pH and calcium : non linear relationship

4.3.1 Managing soil pH

The management of soil pH in cultivated fields has often been approached with too much leniency. Recommendations often advocate for a target pH around 7 or slightly below 7. Furthermore, the cause-and-effect relationship between excess or deficiency of calcium ions and respectively high and low pH levels is consistently made. However, while this is indeed a common situation, it is not universally applicable. Very often, calcium ions are indeed involved in soil alkalinity. But this is not an absolute rule. For some soils, it is potassium ions that contribute to alkalinity, or magnesium ions, or even sodium ions. In these cases, the addition of an external source of calcium can be extremely beneficial because the soils are often deficient due to years of agronomic practices that were not adapted to the situation. Soils near the sea, including polders (sodium), the Limagne region (magnesium), and the Jura meadows heavily fertilized with manure (potassium), are potentially affected by non-calcium alkalinity. In these situations, the addition of a calcium salt will be desirable even if the pH is high. *A contrario*, there are acidic soils that are highly supplied with calcium ions. This is the case for soils formed on gypsum parent rocks ($CaSO_4$), which can be found in Spain, particularly in the Tajo and Elno valleys, in Bulgaria, or in the southwest United States. Gypsum is the calcium salt of sulfuric acid and as such has no neutralizing function. Calcium sulfate in solution has a pH of about 6.5. Therefore, having a relatively low pH does not necessarily mean that calcium needs to be added to rebalance these soils; it may be other bases than calcium that need to be added.

4.3.2 Surface acidification often goes unnoticed

The accurate measurement of pH is another weak point in soil analyses as often practiced. Surface acidifications are often underestimated or simply go unnoticed. Here's an illustration: During an intervention in clayey-sandy loam soils in almond orchard plots in California, we observed extreme pH differences between surface and subsurface horizons. The pH was 4 on the surface and 9 at only 20 cm depth (measured with the Hellige reagent). The flood irrigation method used here for fifty years on these skeletal soils has led to differential sedimentation of elements, with a very high concentration of sand on the surface and clays and silts more concentrated deeper down, accompanied by leaching of surface cations. However, the pH level given by the soil analysis was in this case ideally around 6.8. Yet extremes combined together form a good average, and a plant can die in a good average! In this plot, trees were being uprooted due to lack of productivity, much to the producer's total bewilderment. This illustrates how, solely by reading soil analyses, it is difficult to highlight sometimes major dysfunctions in terms of pH and fertility. In Europe, many farmers, even in regions with carbonate sedimentary bedrock, are facing similar situations often unknowingly, convinced that simply stopping tillage and synthetic chemistry will restore and maintain these mineral and chemical balances.

B

Carbon storage: new pathways

Approaching the soil system by looking at its parts rather than considering it as a whole has led us to adopt an accounting approach to fertility. We have seen how this purely accounting approach is scientifically flawed. This is even more so when it comes to organic matter in soils. Organic matter in soils is primarily a question of balance among the organisms that inhabit the soil. The efficiency of organic matter transformation depends on it. The better the efficiency, the more stable the produced organic matter is. The challenge lies in finding ways to shift the balance and promote the major families of humus builders.

In this section, we'll look in detail at:

- The role of microbiology in the transformation and storage of carbonaceous materials.
- Agricultural practices that contribute to it.
- Natural mechanisms for managing organic materials according to climatic zones and agricultural practices that respect their cycle.

5 • THE IMPORTANCE OF MICROBES

5.1 A paradoxical decline in organic matter levels

The failure to consider the complex processes of soil life has led to several perverse effects in agronomic practices, some of which we have already noted. One of them should have been evident to us for a long time: the continual decline in organic matter levels in agricultural soils since 1950. Despite the increase in biomass yields of cultivated plants (linked to the rise in inputs), the cultivated ecosystem has never returned so much plant residue to the soil, yet our soils have never contained so little humus.

5.1.1 Equation of soil organic matter

Twentieth-century agricultural manuals explained that new methods would trigger a virtuous circle: more fertilizer, higher yields, more organic matter returned to the soil, and hence more humus. The idea that the more you put in, the more you store is now shattered. The humic balance calculations still used today are pure speculation. The isohumic coefficient K1 expresses the yield of organic matter transformed into stable humus, assumed to be constant for each type of material (0.15 for straw, 0.05 for cover crops, for example). However, K1 depends not only on the material, but also on humification conditions. These are average values that we can improve or deteriorate through cultivation practices. As for K2 (the mineralization coefficient), this is also evaluated as a constant. However, K2 is certainly not constant; it depends in particular on cover crops, fertilization and tillage, all of which influence different soil chains. The balance sheet of K1 and K2 represents the setting of living organisms in equations to which these same living organisms have never submitted, much to the despair of a whole generation of agronomists. Why have we arrived at such a paradox?

5.1.2 Everything depends on the microorganisms involved

The fact that organic matter levels are decreasing is clear evidence that the issue of organic matter in soils is not merely a matter of arithmetic between inputs and outputs. The phenomenon is actually the result of agronomic practices that have directed soil life towards mineralization rather than humification. Humus builders—the fungi—are systematically disadvantaged by current practices, including the systematic use of fungicides, deep plowing, green manures resembling lettuce rather than woody material, or inappropriate fertilizer applications that generate salinity. Conversely, bacterial populations, less sensitive to these practices and equipped with powerful mineralization functionalities, have been systematically favored.

5.1.3 New practices still have room for improvement

As long as these mechanisms are not known, understood, and incorporated into agronomic practices, we will continue to steer soils towards mineralization. This holds true regardless of the type of agriculture, even if it appears to be "green." Without any reflection on the subject, organic farming or conservation agriculture still have room for improvement in redirecting soils towards effective and long-term carbon storage. Meta-analyses and joint studies by INRA and Arvalis Institut du végétal (*Effets du travail du sol sur les cycles biogéochimiques (azote et carbone), Effects of soil management on biogeochemical cycles (nitrogen and carbon),* 2014), for example, have shown that no-till practices alter the spatial distribution of organic matter in the soil but not the quantity stored. "The major effect of reduced tillage is to create stratification of organic matter in the profile. The carbon stock is also little or not increased by conservation tillage systems," the study emphasizes. Moreover, returning to plowing after practicing conservation agriculture is not likely to massively reduce carbon stocks in the soil. "Although there are good reasons to use conservation techniques, they do not necessarily lead to carbon sequestration," the study concludes. As for overall greenhouse gas emissions, "the balance is on average little affected by soil management. This is not a determining criterion for the adoption of conservation practices." Mechanization with plowing is therefore not the sole cause of soil degradation! There is no mystery; if we want to restore carbon storage functions in soils, we must first understand the mechanisms behind it.

5.1.4 The pursuit of yield drives us to decapitalize even more

Why are our practices almost systematically unfavorable to carbon sequestration in all systems? Part of the answer lies in the fact that the universally accepted indicator for assessing the value of a practice, the relevance of a route, the efficiency of a fertilizer, or the profitability of a pesticide is yield. Inadvertently but effectively, we have systematically selected everything that decapitalizes. The main capital stored is soil humus. The mineralization of organic matter is a process that provides plants with nutrients in ideal proportions by supplying not only mineral elements but also more complex organic constituents, already partially formed, the basis of living bricks. This process is extremely favorable to plant growth and yield. (We have already mentioned this in part A regarding the transient positive effects of "cides"). Regardless of the type of agriculture, practices contributing to the mineralization of organic matter systematically emerge favorably in agronomic trials. We ourselves conducted trials in situations of weak growth of chlorotic fruit trees. Inputs of metham-sodium, a broad-spectrum biocide molecule with fungicidal, insecticidal, herbicidal, and nematicidal properties, initially lead to remarkable productivity gains, accompanied by the disappearance of nutritional deficiencies: everything stored in the annihilated microfauna and microflora will be instantly made available to the plant. The number of fertilizers, molecules, and cultural operations leading to this "destocking" is impressive: virtually all "cides", improperly calibrated N, P, and K fertilizers, deep plowing, indiscriminate liming, certain green manures, RPEs (rhizosphere priming effects)...

5.1.5 Restoring balance in sequestration and mineralization

However, this decapitalization of soil organic matter is by definition not "sustainable". The negative effects of this depletion are slowly being felt. Beyond the fact that this exhaustion of resources stored in the soil reaches a level that weakens the resilience of soils, crops, and agricultural activity itself, carbon storage in soils now takes on a more universal dimension. It is a promising avenue for reducing greenhouse gas concentrations. In an agriculture aimed at restoring soil potential, the challenge will be to find a balance between sequestration and mineralization.

Achieving net gains in carbon sequestration is entirely achievable and compatible with different production systems. It is possible to effectively store carbon while preserving a favorable and sufficient mineralization capacity. In this section, we aim to provide a better understanding of the mechanisms of organic matter in soils and some strategies to reconnect with a carbon savings dynamic in soils.

5.2 Importance of microorganisms

Soil organic matter content is the result of humification and mineralization processes. It is the microorganisms that drive these changes in organic matter in the soil. In this chapter, we will discuss the importance of different groups of microorganisms in the processes of organic matter construction, or conversely, its availability to plants.

5.2.1 Importance of biological activity in soils cycles

■ **The biomass is entirely connected to the soil**

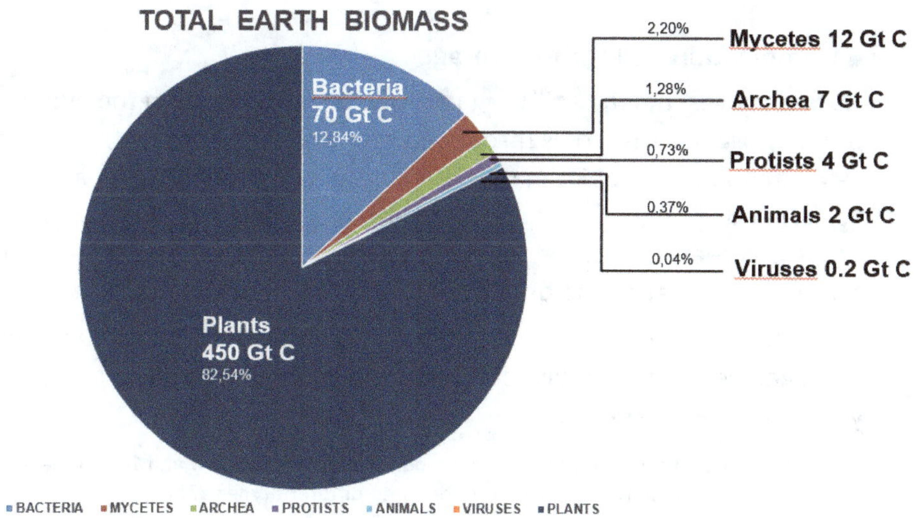

Figure 5.1 – Distribution of terrestrial biomass according to groups of living organisms (per Bar-On et al., 2018)

□ The composition of biomass

The total estimated terrestrial biomass, approximately 600 gigatons (of which only 1.3% is found in the oceans), **is comprised of 82% plants**. Following that are bacteria (12.84%), fungi (2.20%), archaea (1.28%), and finally protists (a group of unspecified unicellular organisms) at 0.73%. Animals ultimately make up only 0.37% of terrestrial biomass.

□ Microorganisms, key players

Microorganisms are ubiquitous in soils. Their mass (almost equivalent to that of roots) and the exchange surfaces they maintain with the soil environment (much larger than roots) are colossal. They largely control the carbon cycle. They are the guardians of carbon-soil-atmosphere exchanges (Schimel and Schaeffer, 2012; Gleixner, 2013). Therefore, they constitute a prominent lever (arguably the foremost one) capable of directing soil function in various directions and influencing the global carbon cycle.

■ Primary production provides the soil's energy ration

The significant biomass in the soil, composed of microorganisms, primarily derives its energy from root exudates and organic residues, mainly of plant origin. These two sources provide microorganisms with "room and board" at a regular pace. Between meals, specialized microbial decomposers go dormant (making way for other assimilative or chemoautotrophic microbes).

■ Humification and carbon storage

Humus represents the soil's organic matter resulting from the evolution of primary organic materials (produced primarily by photosynthesis), as well as animal and microbial materials. Humus consists of humic acids, fulvic acids, and humin (nitrogen-rich organic matter). Various biological processes lead to humus formation, with highly variable efficiencies in terms of the quality and quantity of humus produced.

The decomposition and humification process of crop residues will, on average, yield the following for 1 kg of straw residue (much less favorable yield for green manure).

Table 5.1 – Products from the decomposition of 1 kg of straw and 1 kg of green manure and transformation into different types of humus

	CO_2 respiration-metabolism	Labile humus	Stable humus	Coefficient K1
Straw	600 to 800 g	50 to 120 g	20 to 60 g	0.15
Green manure	800 to 950 g	15 to 50 g	5 to 15 g	0.05

□ Mineral balance of humus

Humus is primarily composed of C (carbon), N (nitrogen), S (sulfur), and P (phosphorus) in the following proportions: 100 parts carbon / 10 parts nitrogen / 1 part sulfur / 1 part phosphorus / 1 part potassium / + 0.1 trace elements. Based on 3,000 tons of soil per hectare (see table), every percent of organic matter in the soil represents 30 tons of matter per hectare. This equates to 3,000 kg of nitrogen, 300 kg of sulfur, 300 kg of phosphorus, and 30 kg of trace elements. The rate of mineralization varies greatly (from 0.5 to 3% per year). The significance of this composition lies in its origin from living organisms; humus is a concentrate of life, and its mineralization enriches the soil solution with a cocktail that fully meets the needs of crops.

Table 5.2 – Mineral richness represented by one point of organic matter per hectare (based on 3,000 tons of soil per hectare)

SOM in kg/ha	C	N	S	P	Oligos
30,000 kg	27,000 kg	3,000 kg	330 kg	330 kg	30 kg

Soils sequester more than half of the Earth's carbon

Carbon entering the carbon cycle is primarily present on Earth in the form of organic matter. The living terrestrial biosphere contains roughly the same amount (620 GtC) of carbon as the atmosphere. Organic matter in soils, composed of humified and evolving plant and animal tissues, along with the living microbial community and macrofauna, contains more than twice as much carbon, totaling approximately 1,580 GtC (Gleixner et al., 2001).

□ The mineralization coefficient (K2) is actually highly variable

Under conditions of minimal soil disturbance, these stored elements will be mineralized very gradually. However, with energetic soil cultivation, much larger quantities will be mineralized due to increased oxygenation. Depending on the management practices, the mineralization, and the depletion of the organic matter stock will occur at a rate ranging from 0.5 to 3% per year (K2). At the lower end of the range, we will often encounter problematic accumulations of unevolved organic matter (such as rankers, peatlands, or cold soils with excessive slurry/manure). Conversely, at the higher end of activity, in good loams supporting crops requiring intensive soil cultivation (such as sugar beets, potatoes, vegetables), we will experience rapid decreases in humus levels. Accelerated mineralization results in abundant harvests but also in significant waste of the carbon and nitrogen stored in the soil organic matter. This is what happens when a pasture is plowed under. The variability of K2 is one of the factors, among many others, that will always challenge even the most accurate predictive humic balances.

□ **Simply considering the quantities of litter added is not sufficient**

Today, predictive models of organic matter evolution are still primarily based on the amounts of litter added. However, in reality, the dynamics of soil organic matter are mainly under microbiological control. Therefore, the key to humus formation lies not so much in the quantity of organic matter returned to the soil, but rather in the activity of certain groups of microorganisms. Knowing this, we can assume that the rate of carbon sequestration in soils likely responds much more rapidly to changes in climate and vegetation than previously thought. However, these hypotheses are still not integrated into climate change models (Detlef, 2005), and they are rarely incorporated into agricultural practices.

5.2.2 Microorganisms: major players in humus dynamics

Microorganisms are the major players in the organic matter cycle in soils. Here is an overview of these various actors and their ability, or lack thereof, to build large quantities of stable and functional humus.

■ **Fungi**

1. **Fungi** play important roles in water circulation, nutrient cycling, carbon cycling, and disease regulation. Soil fungi can be grouped into three general functional groups based on their feeding mode.
2. **Saprophytic fungi**, including basidiomycetes (mushrooms) and ascomycetes, convert dead organic matter into fungal biomass, carbon dioxide (CO_2), and small molecules such as organic acids (some of these fungi are pathogenic). They typically utilize complex substrates such as cellulose and lignin. They may also contribute to the decomposition of natural or synthetic organic pollutants (such as pesticides). They help increase the accumulation of organic matter rich in humic acids, which resist degradation and can persist in the soil for hundreds of years.
3. **Mycorrhizal symbiotic fungi** colonize plant roots. In exchange for plant-produced sugars, mycorrhizal fungi help solubilize phosphorus, mineralize humus (with the assistance of associated microflora), and provide plants with soil nutrients (phosphorus, nitrogen, trace elements, and possibly water).
4. **Pathogenic or parasitic fungi** (some of which are also saprophytic). Airborne pathogenic fungi (such as mildew, fusarium, septoria, etc.) and root pathogens, such as *Verticillium*, *Pythium* and *Rhizoctonia*, are essentially "opportunists" that take advantage of plant weaknesses. They act as ecosystem cleaners.

□ **Strict aerobes**

Fungi are strict aerobic organisms; they can only thrive in the presence of oxygen. Soil fungi are thus very sensitive to oxygen deficiency, compaction, and waterlogging. This is why observing soil structure is crucial (see "soil profile"). Neither saprophytic fungi nor mycorrhizae appreciate fungicides. They prefer slightly acidic to acidic pH levels, underscoring the importance of proper use of amendments and fertilizers (see "Fertilization" chapter). They are highly sensitive to soil cultivation practices that disrupt their mycelial networks.

□ **Ecology of saprophytes: organisms favored by woody materials**

Saprophytes (basidiomycetes and ascomycetes) thrive when provided with substrates rich in cellulose and lignin. They gain a competitive advantage over bacteria when straw is returned to the soil, but the opposite is true when green manures are returned. "Decomposer" fungi are remarkably efficient at utilizing and storing carbon (40 to 55% of the original carbon) and have a carbon-rich C/N ratio ranging from 10:1 to 25:1. The lignin content is much more decisive for degradation speed than the C/N ratio of plant residues (lignin is a complex compound and takes longer to digest). Moreover, a high lignin content is much more decisive in generating a preferential development of fungi. Indeed, only fungi are equipped with the enzymatic arsenal for this operation and are thus capable of releasing the cellulose bound to it. Whenever woody material rich in lignin is added to the soil, fungi and humus builders are favored. Whenever a ration rich in soluble elements and cellulose is added, the functioning of bacterial chains is favored.

□ **A ration more or less suitable for fungi**

The digestibility of forages by cows is a well-documented example that illustrates how bacteria prefer materials rich in simple sugars and cellulose and how they are poorly equipped to digest lignin. The cow's rumen is a true cellulose-degrading factory. The rumen is primarily colonized by cellulolytic bacteria and few, if any, fungi. The digestibility of a forage by the cow is maximal (95-100%) during the first cycle of forage when the lignin content is nil. The digestibility of forages by cows decreases with the age of the forages, directly related to the proportion of lignin. Furthermore, we will see later that the ecological role of ruminants in ecosystems is to impoverish the ration returned to the soil in terms of cellulose and soluble elements (in the form of dung).

The construction of stable humus in soils involves the return of organic matter that is either senescent or digested by ruminants.

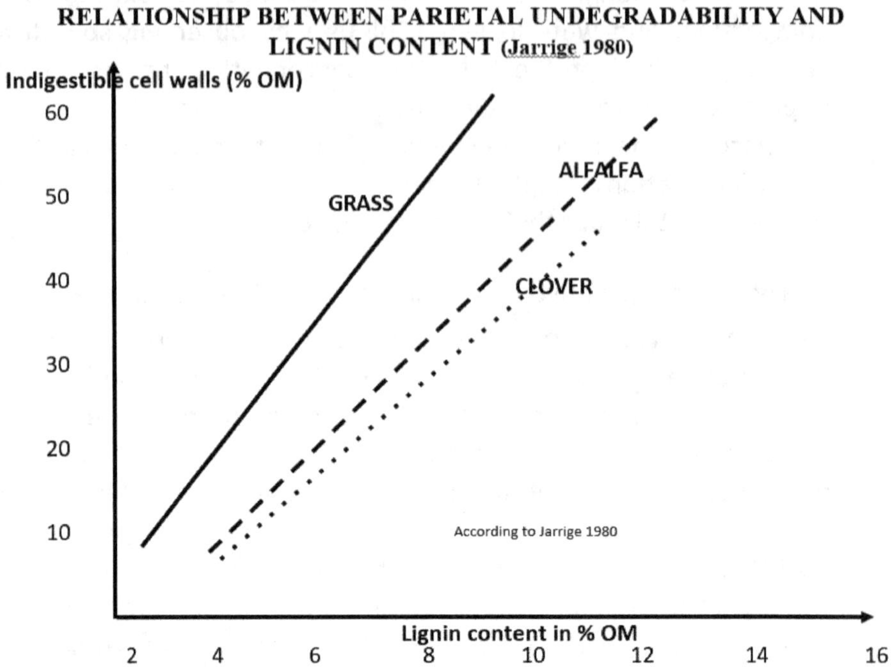

Figure 5.2 – Relationship between indigestible cell walls and lignin content (as % of OM, Jarrige, 1980)

□ Ecology of mycorrhizae

Mycorrhizae do not establish themselves on Brassicaceae or Chenopodiaceae (such as cabbage, rapeseed, mustard, beets, spinach, etc.). Their numbers notably decrease in the presence of these botanical families in crop rotations. They cannot tolerate mineral imbalances, especially excesses in terms of calcium-magnesium ratios, phosphorus content, and nitrogen content, which we will discuss in the section dedicated to fertilization. Frequent and deep soil tillage is harmful to them, as is the use of broad-spectrum fungicides.

Mycorrhizae develop hyphae 20 to 50 times finer than plant roots, on the order of micrometers (µ), equivalent in size to clays. This characteristic allows them to explore places inaccessible to plant roots and provide them with minerals, phosphorus, nitrogen, and water that would otherwise go unutilized. It's a win-win exchange with the host plant, which provides carbohydrates to the fungus: both sugars and lipids. The plant greatly benefits from this exchange. It shares its sugars, but gains in "net energy" by accessing minerals at a lower cost: the extension of its roots by hyphae, compared to roots, allows it to save enormous amounts of carbohydrates. These mycelia contribute to structural stability, prevent soil particles from aggregating, and even prevent or repair compaction.

■ Bacteria

Bacteria are much smaller than fungi and well adapted to harsh conditions and impoverished environments. Drought, flooding, intense soil cultivation... Their populations survive extreme conditions because their small size allows some colonies to find microsites in the soil that are less extreme.

Bacteria are rich in amino acids and organelles, containing 10 to 30% nitrogen. They are less efficient than fungi at converting original carbon from plant primary production into organic molecules. As soon as conditions become favorable, they multiply rapidly, with their population doubling in 20 minutes. When conditions are unfavorable, facultative anaerobes, which are numerous, use a crisis metabolism that allows them to survive. Under these conditions, strict anaerobes thrive. However, anaerobic activity is not efficient in terms of utilizing the original carbon. While bacteria exhibit 5 to 15% efficiency in carbon transformation in oxygenated environments, this performance drops to 2 or 3%; for comparison, fungi efficiency can reach up to 50%. In such conditions, bacteria leave behind poor-quality humus, with low CEC (cation exchange capacity), characteristic of waterlogged soils, accompanied by methane and carbon dioxide production. Moreover, many anaerobic bacteria are pathogenic and thrive in these conditions.

Bacteria can be classified into four main categories:

- **Decomposers**: They contribute to the breakdown of dead organic matter, incorporating it into the humus cycle and plant nutrition. They digest simple materials such as sugars, starch, and lipids with relatively low efficiency (losing 75 to 85% of the original carbon through respiration).
- **Symbiotic bacteria**: The most remarkable are those that fix nitrogen from the air. Some are free-living, such as Azotobacter and *Azospirillum*, etc. while others are in close symbiosis with plants, such as *Rhizobium* (for legumes) and *Frankia*. They exchange their services for carbohydrates with plants. Actinomycetes, once classified as fungi, prepare the ground for fungi but are not as efficient in lignin decomposition. Actinomycetes are thermophilic organisms that prefer neutral to alkaline pH. They are responsible for the earthy smell of soils and increase the temperature of compost piles.

Characteristics	Fungi	Bacteria
Respiration – metabolism	Strict aerobe	Facultative aerobe or strict anaerobe
Impact of soil tillage	Sensitive	Stimulation
No-till	Very favorable	Indifferent
Lignin utilization	Yes	No
Cellulose utilization	Yes	Yes
Preferred OM C/N	C/N > 30 up to 1000	C/N 6 to 25
Recycling efficiency of original carbon C	40-55% Impossible under anaerobic conditions	5-15% in aerobic conditions 2-3% in anaerobic conditions
Excess P, N effect	Inhibition	Stimulation
Effect of compaction	Inhibition	Metabolism modification
Effect of hydromorphic soil	Inhibition/Destruction	Metabolism modification
Fungicide effect	Highly sensitive	Indifferent
C/N ratio of tissues	10-25	5-7
Nitrogen requirements	Min 0.2%	Min 2%
Preferred pH	5 to 6.8	5 to 8
Mineral and water transport	Up to 15/20 cm	0
Improvement of soil structure	+++	+/-

Table 5.3 – Comparative characteristics of fungi and bacteria

- **Pathogens**: *Erwinia, Pseudomonas*... These opportunistic organisms, known as "ecosystem cleaners," fully illustrate the principles of trophobiosis.
- **Lithotrophs**: They derive their energy directly from minerals as electron acceptors. They include sulfur bacteria, *Nitrosomonas*, iron bacteria, and anammox bacteria. In soils, they actively participate in clay formation when in contact with the parent rock.

■ Protozoa

Protozoa are significant consumers of bacteria. Their population evolves alongside that of bacteria. One of their essential functions in soil is to regulate bacterial populations while releasing nitrogen in the form of available ammonium for plants. Protozoa also have the capacity to store nitrogen in the autumn/winter periods in order to prevent leaching.

■ **Large families with different ecological needs, providing leverage for action**

To better manage carbon storage, it can be useful to visualize the characteristics of the two major families of soil organisms below. These characteristics dictate the strategies of fertilization, soil management, and cover crop management that we will propose.

5.2.3 Importance of exudates and rhizodeposition

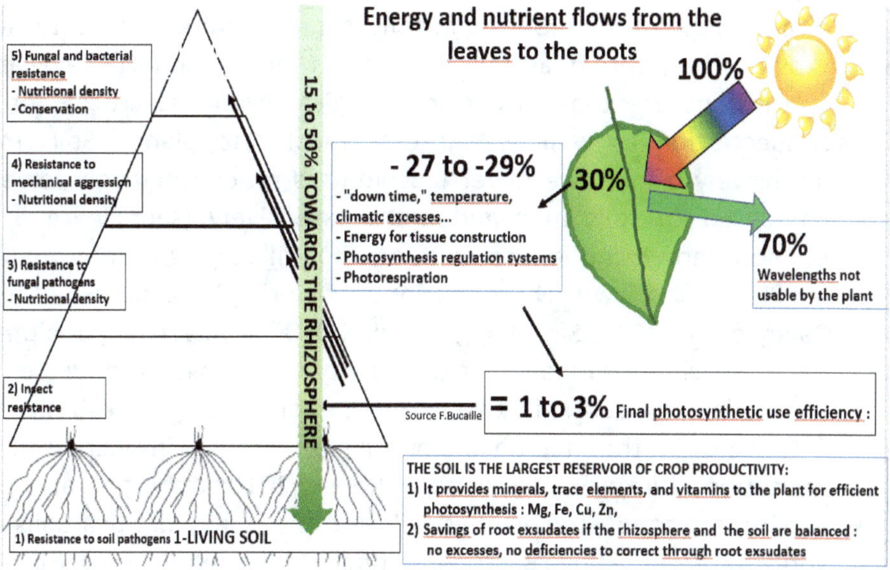

Figure 5.3 – Flow of carbohydrates from leaves to roots and vice versa

Plants "invest" in the soil through their root exudates. Through these exudates, they are capable of shaping the rhizospheric microbiology through the transfer mechanism described below. These exudates nourish the microflora (fungi, bacteria as described earlier). This is yet another mutually beneficial exchange between the plant and these microscopic helpers. Indeed, the plant produces simple sugars (glucose) through photosynthesis, which it releases to the microorganisms. These microorganisms then metabolize (construct) them into complex molecules that they leave in the soil (as humus) or directly provide to the plants. The energy and nutritional flows go both ways. The soil and rhizosphere provide quality to the plant: minerals, trace elements, vitamins, and water through upward sap flow. But the plant also invests, through downward sap flow, 15 to 50% of its photosynthates to ensure the development of its rhizospheric microflora. The plant, when compelled to modify an unfavorable root environment, does so through root exudates, whose action is direct due to their composition and indirect through the orientation of microbiology. The more extensive this rebalancing, the more the plant wastes its resources. Hence the importance of soil rebalancing as described in section E.

5.2.4 The bacteria/fungi ratio is decisive

The ratio between fungi and bacteria (F:B for Fungi:Bacteria) is decisive in the evolution of soil organic matter (Johnson, 2015). Numerous studies support this. Dr. David Johnson, based at the University of New Mexico, discovered that the F:B ratio in soils is a more important factor for plant growth than the quantity of nitrogen or phosphorus available. His conclusions are that this microbiological profile is crucial for carbon sequestration as well as nutrient transfer to plants. Soils managed intensively often have lower F:B biomass ratios (relatively fewer fungi) compared to soils managed more extensively (see Figure 5.4). This phenomenon is thought to be due to soil tillage, significant quantities of fertilizers, and low C/N organic inputs favoring bacteria (Six et al., 2006; Bailey et al., 2002; Sinsabaugh et al., 2013). A lower fungal biomass has been associated with a lesser capacity of these soils to sequester carbon. It is believed that a shift towards fungal dominance, which has a higher C/N ratio in the microbial community, results in increased carbon utilization efficiency (Strickland et Rousk, 2010; Waring et al., 2013). An experiment conducted on physically and chemically identical soils but with very different Fungi/Bacteria ratios (Malik et al., 2016), involved measuring the loss of carbon 13 (^{13}C) after incorporating *chenopodium ambrosoides*. After one month, the carbon loss through microbial respiration was 44% for the fungi-dominant condition, compared to 65% for the bacteria-dominant condition. The amount of carbon incorporated into the soil organic matter was respectively 53% for the fungi and 32% for the bacteria-dominant soil.

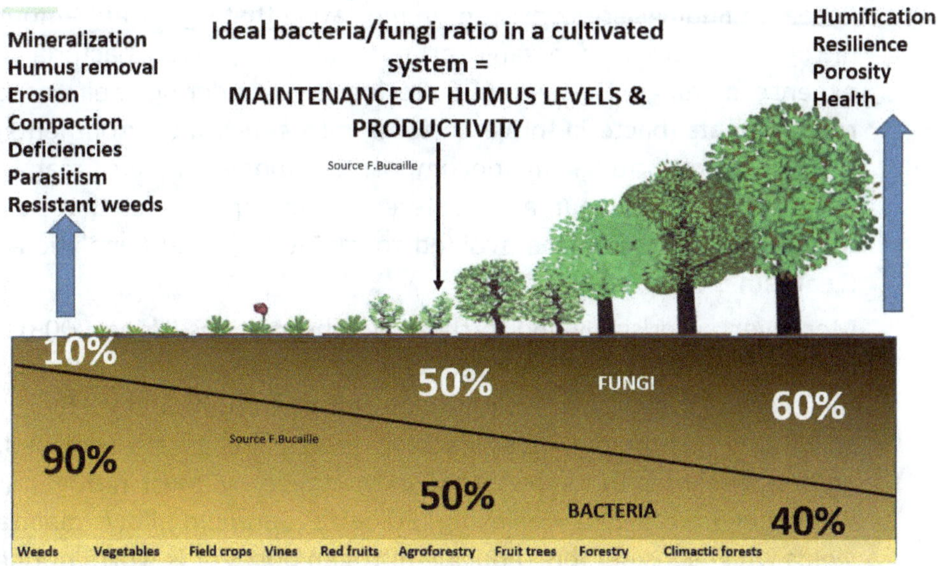

Figure 5.4 – Fungi:Bacteria ratio by farming method and plant cover

In addition to being more efficient in building humus in soils, fungi create organic materials that are also much more durable. The half-lives of fungal materials are incomparably higher than those of bacteria: six months to three years for bacteria and 20 to 50 years, or even more for certain fungal-origin components (glomalin, melanin)! Furthermore, considering their own C/N ratio, fungi are more inclined towards carbon storage than bacteria. Bacteria, on the other hand - with a lower C/N ratio - have stronger nitrogen requirements to store the same amount of carbon.

Saprophytic fungi better utilize primary organic production and for a longer period. With the same amount of material, they produce more humus, which is also more durable and of higher quality. Therefore, they are the key lever for long-term carbon storage in soils.

Glomalin: a major structural role

Glomalin, a glycoprotein produced by mycorrhizal fungi, has a significant impact on structural stability by aggregating minerals and humus, thereby increasing the Cation Exchange Capacity (CEC) of soils. Unlike other humic compounds, this molecule is directly produced by the fungus and serves to protect the fungus against bacterial attacks. Additionally, possessing hydrophobic qualities, this molecule appears to be strongly correlated with structural stability and carbon sequestration in soils (Rillig et al., 1999). As previously mentioned, soil compaction or aeration is primarily under biological and microbiological control. "In the soil, roots and microbes (especially fungi) are the carpenters who maintain the house, continuously producing the glues (polysaccharides and glomalin) that hold the parts together" (Hoorman et al., 2011).

5.2.5 Fungi are penalized by nitrogen

Research findings are consistent in indicating that significant amounts of nitrogen, whether in organic or mineral form, favor bacteria at the expense of fungi. This seems quite logical considering their respective requirements (bacteria thrive better in nitrogen-rich environments). The most effective fungi in decomposing woody organic matter are basidiomycetes, and numerous references support the idea that excessive nitrogen levels should be avoided to maintain fungal lignolytic activity (Leatham & Kirk, 1983).

Moreover, surprisingly and fortunately, observations (Vries, 2009) reveal that once nitrogen levels are reduced sufficiently to allow fungi to regain their ecological niche entirely, the nitrogen utilization efficiency is significantly improved, and leaching losses are greatly reduced. The agroecosystem is therefore not disadvantaged by lower nitrogen inputs when fungi are active. Achieving biological optimum allows maintaining yields while considerably reducing nitrogen doses (it is worth noting that nitrogen efficiency is very low in conventional systems, around 35%).

6 • NITROGEN STARVATION: MYTH AND REALITY

For several decades, we have observed in the field that the decomposition of organic matter is slow and often detrimental to the crops in place. Two mechanisms are generally invoked. The first is nitrogen deficiency, which is real but rarely the primary cause. The second, seldom mentioned, corresponds to allelopathic phenomena. In this chapter, we will explore why the notion of nitrogen deficiency is overestimated and mostly indicates a hunger for fungi.

6.1 Nitrogen alone is not a limiting factor in decomposition

6.1.1 Talk of nitrogen starvation is often questionable

The “nitrogen starvation” mechanism implies that nitrogen is often the limiting factor in the decomposition processes of organic matter and results in a lack of available nitrogen for the crops in place. It is true that the microorganisms involved in the decomposition of plant organic matter have a relatively low carbon-to-nitrogen ratio (they are nitrogen-rich), ranging from 10 to 16 for *aspergillus*, and around 14 for bacteria (Mangenot, 1980). These organisms indeed concentrate nitrogen in their cell walls through their enzymatic arsenal, which alone represents 40% of the dry matter of a bacterium! Microbial syntheses thus consume a lot of nitrogen. *A contrario*, the carbon-to-nitrogen ratio of litter and crop residues is high (from 80 to over 100 for straw and ramial wood chips). Hence, the hasty conclusion that the C/N ratios of the materials added are unfavorable and induce mobilization of nitrogen from the soil by the decomposition flora. This has led to the notion of a "nitrogen starvation" in the soil, which should be satisfied by supplying mineral nitrogen.

6.1.2 No nitrogen starvation in forests

In “natural” forests, there is no "nitrogen starvation" yet the organic matter returns to the soil are even poorer in nitrogen. This is because they consist of entire trunks, of which 90% of the mass is dead (with C/N ratios of 400 and above, up to 1,000), and dead leaves (C/N ratios of 60 to 80), rather than green manure (C/N ratios of 10 to 30), straw, or ramial wood chips (with C/N ratios of 80 to 120). Forest humus is commonly poorer in nitrogen (C/N around 15) than cultivated soils (C/N 8 to 12). Philippe Duchaufour cites in *La revue forestière (Forestry review)* C/N ratios for litter under pine and fir trees of 20/23, and under beech and broad-leaved trees of 15/17 (Duchaufour, 2015). Despite the slightly higher C/N ratios in nitrogen-poor forest humus, we can observe that the virtuous process of decomposing nitrogen-poor litter into evolved humus is entirely efficient without the addition of mineral nitrogen. The higher C/N ratio reflects the composition of the actors involved: fungi, but also the ability of mycorrhizal fungi to extract nitrogen from fresh organic matter and litter and transfer it directly to growing plants (Hodge, 2001). In these forest situations, we have never observed "nitrogen starvation"; quite the opposite. Some important evolutionary decomposition processes are repressed by nitrogen richness, both through enzymatic repression (Reid, 1979 or Commanday et al., 1985) and "as a result of the stimulation of microbial antagonisms" explains F. Mangenot (1980). Too much nitrogen impairs the expression of the lignolytic function of soils, carried out by fungi. The nitrogen content thus acts as a regulator of soil microbiota. This results in maintaining a balance between decomposition and assimilation of organic matter, between humification and mineralization. We can see that nature has the ability to assimilate organic matter quite low in nitrogen and rich in cellulose and lignin without external inputs of mineral nitrogen. If the agroecosystem does not show an equivalent capacity, it is primarily because it has lost these functionalities due to the breakdown of basic trophic chains. **Instead of invoking "nitrogen starvation" as an assumption, let us rather ask ourselves why decomposition functionalities are currently inactive and how they could be restored.**

6.1.3 Fungi starvation

The nitrogen starvation often described actually stems more from fungi starvation in the soils. Indeed, the mechanisms of organic matter decomposition and carbon storage in the soil rely on efficient enzymatic hydrolysis of celluloses and lignins. The most active players in this regard are fungi, especially Ascomycetes and Basidiomycetes. It is no coincidence that forests harbor so many of them, especially in autumn when dead branches and leaves fall to the ground. Basidiomycetes are equipped with enzymes (ligninases, laccases, cellulases) essential for ensuring rapid breakdown of massive inputs of raw materials.

The arrival of white rot fungi: a "technological leap" marking the end of the Carboniferous era

Relatively recent research (Floudas et al., 2012) provides further evidence of the ecological contribution of fungi. The Carboniferous period lasted 60 million years, from 360 to 300 million years ago. This period was characterized by a steady, dangerous decline in the level of atmospheric CO_2. If this decline had continued, photosynthesis and other environmental evolutionary processes would have arrived at a standstill. Nature was at a dead-end: carbon had to be recycled. It ended when white rot fungi, armed with lignolytic and cellulolytic enzymes, appeared. Before that, fungi existed but were not equipped to perform the recycling of dead organic matter. It was with their appearance that the accumulation of peat transformed into coal ceased. The emergence of these fungi represented a technological leap in ecosystems, unlocking a barrier in the decomposition of organic matter and ushering in a new era.

6.2 How the "miracle" of decomposition by fungi works

6.2.1 Very high C/N ratios: the playground of lignivorous fungi

The nitrogenous materials contained within these recalcitrant materials need to be released from their lignocellulosic and especially woody matrix to be made available; thus sustaining the nitrogen chain in the food web involved in the decomposition process of materials. This is what fungi excel at. It is noteworthy that basidiomycetes, in forest environments, operate within even more unfavorable ranges of nitrogen and carbon sources than in ramial chipped wood, straw, and other mature crop residues. The C/N ratios of forest litter range from 170 to 200 and beyond. Fungi decomposing woody materials can thrive on woody materials containing as little as 0.03% to 1% nitrogen with C/N ratios that can exceed 1,000:1. Additional nitrogen input thus does not seem to be an indispensable condition for the humification of cellulosic or woody materials. However, their mycelia contain on average 5% nitrogen with a C/N ratio of 10:1, and the humus they create is rich at a ratio of 15:1.

6.2.2 Nitrogen relocation by fungi

How is such intense activity possible on materials so low in nitrogen? Lignivorous fungi have several mechanisms for optimizing the nitrogen cycle. They can relocate nitrogen from older hyphae to those that are actively growing. They can also preferentially allocate the limited nitrogen available to vital functions. *Coriolus Versicolor* (a lignivorous basidiomycete) has been studied in environments with both low and very high C/N ratios (Levi & Cowling, 1969). The nitrogen percentage in the hyphae dropped from 4.4% (in nitrogen-rich conditions) to 0.2% (nitrogen-poor conditions). However, simultaneously, the nitrogen percentage in nucleic acids increased from 4% to 25%, and notably, the quantity of cellulase produced per unit of mycelium remained the same! White mycelium fungi can produce their cellulases even in C/N ratios as high as 2000:1.

6.3 Decomposition by fungi can be promoted

6.3.1 When fungi can no longer "work"

Lignolytic fungi operate optimally when metabolizable carbon compounds (sugars) and available nitrogen are at their minimum, and the oxygen concentration is at its maximum. This characteristic gives fungi a decisive ecological advantage by granting them access to celluloses and hemicelluloses embedded in lignin. Therefore, it is certain that green manure will not give them much opportunity to thrive. However, other factors limit their action:

1. **Natural antifungal substances** (terpenes, resins, tannins, polyphenols). This is an issue in spruce forests where organic matter turnover is very slow. It is also a problem with wood chip mulch (RCW) derived from such forests.
2. **Synthetic antifungals** (synthetic or mineral fungicides). This is a problem in cultivated agricultural areas. Consequently, the origin of the materials returned to the soil, how they have been treated, and the presence of various residual compounds in the soil are all crucial factors.

Cereal straw treated with SDHI (Succinate DeHydrogenase Inhibitor), strobilurins, or copper sprays will exhibit significantly impaired transformation coefficients and rates due to the inhibition of decomposing fungi. Under these conditions, nitrogen starvation can occur. In such scenarios, the task of organic matter decomposition is "entrusted" to actors other than fungi, such as bacteria. These bacteria have significantly higher nitrogen requirements but can adapt to more severe soil conditions. High sugar content in young covercrops, significant salinity (from mineral fertilizers), and the presence of fungicides definitively favor bacteria over fungi (see Table 5.3). In these conditions, nitrogen starvation occurs because bacteria have a much more robust metabolism than plants and will monopolize all available nitrogen.

6.3.2 Nitrogen starvation is also a consequence of certain practices

Nitrogen starvation is a consequence of agricultural practices. Typically, those who invoke this concept suggest that soils are necessarily malfunctioning, with deficient trophic chains in managing mature materials. While this situation is indeed common in current agroecosystems, it is certainly not normal. Nature is capable of humifying very low-nitrogen materials provided that the conditions allow it. We can restore this functionality by favoring fungi. Any practice aimed at restoring fertility by increasing organic matter in soils should take into account the plant species, the origin, and the technical production process of these materials within the framework of an agroecosystem.

6.3.3 Revealing trials stemming from the implementation of this knowledge

Trials (see table below) conducted over two years (2011 and 2012), with three replicates on strips of 24 × 400 m (a 35 ha plot), clearly highlighted yield differences depending on the management of the crops originating from the spread straw and the methods of straw restitution to the soil. Yields were visibly penalized under conditions of slow straw degradation linked to a significant "load" of fungicides. Fungicides inhibit fungal flora involved in organic matter decomposition, leaving the field open to nitrogen-hungry bacteria. Their application at the end of the cycle on plants intended to be returned to the soil should be absolutely prohibited. Finally, zinc, manganese, copper and a number of other trace elements are crucial to the performance of cellulase and ligninase, and must be present at a verified minimum level.

When there is doubt about the possible presence of a nitrogen starvation mechanism, it is always good to check nitrogen residues or, better yet, to use the rapeseed cress test (Afnor FD U44-165 standard, December 2015). Indeed, the seeds of this plant are sensitive to all allelopathic substances inhibiting germination and growth; what is measured is the seedling emergence rate, and at this early stage, obviously, nitrogen deficiency cannot be the cause of germination failure. Often, this allows us to verify that the observed problem does not stem from nitrogen starvation and thus save on input while focusing on the core issue.

Table 6.1 – The fungicide load from the previous crop penalizes yields and the evolution of cellulosic materials (source: trials conducted in Nièvre (France) over two consecutive years, 2011 and 2012, with three repetitions of 24 x 400 m on a 35 ha plot).

	Rapeseed in decitons	Wheat in decitons
T1 straw removed (previous crop: barley)	38.1	68
T2 straw shredded from the previous crop (barley without fungicide, with SDE)	42	73
T3 straw shredded from previous crop (barley with strobilurin/SDHI)	36.8	68
T4 compost 10 T of goat manure (made with wheat straw without fungicide). Previous straw removed	43.3	78.1
T5 compost 10 T of goat manure (made with wheat straw + strobilurin/SDHI). Previous straw removed	38.6	70.2

6.3.4 No nitrogen starvation on surface residues

When organic materials are left on the surface, there is no reason to suspect the establishment of a nitrogen starvation mechanism. Indeed, there is no mechanism for deep nitrogen recovery operated by the decomposition flora (apart from fungal hyphae). The reorganization of available nitrogen occurs in the immediate environment of the bacterial mucilage operating on plant residues. Furthermore, there is abundant scientific literature documenting the activity of a free nitrogen-fixing flora that collaborates with decomposing fungi when lignocellulosic materials are left on the surface. Under these conditions, when a vegetation deficiency is observed, there are multiple other avenues to explore.

6.4 Antagonisms between assimilation and decomposition flora

Cultural practices, especially fungicides, are not the sole cause of the mechanisms often described as "nitrogen starvation." The competition between decomposition flora (those that decompose fresh organic materials) and assimilation flora (those that develop around roots, the rhizosphere) also plays a major role. H.P. Rusch in his book « *Bodenfruchtbarkeit* » (« *Fertility of the ground* »), published by K. F. Haug Verlag. (1968) thus highlights the antagonism between decomposition flora and assimilation flora. A soil volume that hosts a decomposition process cannot simultaneously function in the assimilation direction. Moreover, decomposition flora are always more competitive than assimilation flora. As long as the former are still very active, plants cannot perform well. All crops planted very early after the harvest of cereals with straw are thus very sensitive to this competition effect. However, this lack of vigor and nutrition in the young plant is rarely caused by nitrogen starvation. The most common case is the cultivation of rapeseed, where the intercrop period is only a few weeks between the harvest of the previous crop (usually barley) and seeding. From our own experiences, even with nitrogen supplementation in these situations—even ammonium sulfate or ammonium diphosphate—the nutritional state of young plants does not improve. It appears that neither sulfur nor nitrogen (both involved in protein synthesis), nor phosphorus (involved in energy processes) can overcome this visible vegetative poverty. (In these situations, conducting the rapeseed cress test is sufficient to convince oneself.) Competition between flora explains plant behavior, but certainly not in all situations.

6.5 Allelopathic substances at play

The growth defects of a young seedling following the incorporation of mature materials also relate to another phenomenon: that of allelopathic substances contained in the tissues of the previous crop. Polygonaceae, including wild knotweeds and buckwheat, many thiol-releasing crucifers, and especially straw cereals containing significant amounts of coumarin, generate imperfect germination and slowed growth of the subsequent crop. Initially produced to protect living tissues, residual coumarin continues to produce its protective and inhibitory effects post-maturity and after return to the soil. Rapeseed following cereal will inevitably suffer. Conversely, direct wheat sowing after corn grain harvest very rarely shows growth defects. There may be up to 10 tons of corn stalks with a C/N ratio of 120; the same observation applies to sunflowers. Yet all conditions are met for a supposed "nitrogen starvation" to manifest. If viewed from the perspective of allelopathy, it can be noted that corn and sunflower do not contain coumarin. We know from literature and experience the potency of these molecules: couch grass root exudates exert a powerful inhibitory effect (Lastukva, 1955) on the germination of rapeseed, mustard, red clover, barley, and vetch, and even cause the death of wheat. We have even regularly observed that dense quackgrass (*Agropyron repens*) destroyed by glyphosate "sterilizes" the soil to the point of preventing any vegetation from establishing itself for several years (Grümmer, 1961). It releases significant quantities of poisonous exudates before dying. Another concomitant mechanism that can be triggered is the decomposition microflora (*Aspergillus* and *Penicillium*) of straw producing a substance, patulin, highly inhibitory to growth by degrading root absorption processes. This phenomenon is much more related to cultural systems that are completely out of sync with climatic conditions (see § 9.2), as the soils are still warm and rich in mineral residues at the time of harvest. Therefore, corn, sunflower, and sorghum are not affected by this phenomenon, unlike straw cereals.

Carlos Crovetto's windrows

Straw, RCW... Various organic materials of plant origin returned to the soil contain allelopathic compounds. In other words, they have a negative effect on the development of other organisms or plants. Cereal straw in particular contains an allelopathic compound known as coumarin. Chilean farmer Carlos Crovetto (cf. *Les fondements d'une agriculture durable* (*The foundations of sustainable agriculture*), 2000) has observed on his own farm that straw allelopathy begins as soon as the first rains fall, during the initial stages of organic matter decomposition. Moreover, it takes between 60 and 100 mm of precipitation before the straw is washed of its coumarin. Having realized this, he developed a special straw management technique. At each harvest (year N), he gathers the straw into windrows, which he leaves to mature on site for a whole season before spreading them over the whole ground in year N + 1. The 3% loss in surface area associated with this practice is offset by an average yield increase of 6-8%. According to the Chilean farmer, allelopathy no longer produces phytotoxic effects at soil temperatures below 4°C and above 15°C.

7 • GREEN MANURES DO NOT DURABLY STORE CARBON IN SOILS

It is common to hear that the richest organic matter environments are forests and permanent pastures, which is true. And it is often added that green manures function like pastures and therefore will lead us to similar results. The very big difference is that all the herbage production of a pasture is first intended to be ingested and then digested by a ruminant herbivore. In this chapter, we will see that "green manures" are a double-edged tool for carbon storage, and we will provide ways to optimize the management of intercrops or cover crops.

7.1 Green manure is not grassland: ruminants are missing

The bulk of the primary production of a permanent pasture is intended to be first ingested and then digested by a ruminant herbivore and returned to the soil in the form of excreta rich in lignin but poor in soluble elements. Their recycling involves their burial in the soil by coprophagous insects. Through their mechanical action, by creating galleries, dung beetles improve macroporosity and prevent compaction phenomena, thus improving the physical properties of the soil: texture, infiltration rate, porosity. Dung is the undigested fraction of the diet, mainly composed of lignified tissue fragments, cell walls, and bacteria, dead or alive, originating mostly from the rumen and representing 10 to 20% of the weight of feces. Jarrige (1966) analyzed 92 samples of feces from ruminants ingesting green forages and found no soluble carbohydrates. The other constituents were 87% nitrogen-rich organic matter consisting of 80% bacterial nitrogen, cellulose associated with lignin, virtually indigestible chlorophylls, as well as all lipids contained in plant cuticle waxes, all indigestible by bacteria (Grenet, 1970; Akin, 1979).

7.1.1 A bacterial incubator

Rumen is a bacterial incubator that digests everything pleasing to bacteria and leaves indigestibles to the soil fungi, which are their preferred food. All these "indigestibles" are precursors of humic acids and stable humus. As soon as lignin is present, it prevents bacteria from accessing the bound cellulose. Thus, lignin protects three times its weight of hemicellulose and cellulose from bacterial degradation. As we have seen (see Figure 5.2), herbaceous forages, as they mature, accumulate lignin and quickly lose their digestibility. This is understandable since ruminants do not have fungi in the rumen that could perform this work. Digestibility by ruminants (i.e., by rumen bacteria) is high (95%) on young tissues and drops to 50% at the flowering stage, and further decreases later on. Therefore, no, grazed grass, even when young, does not at all produce the same result on soil microbiology as a green manure that is shredded or destroyed by glyphosate (which additionally inhibits soil fungi). Besides the absence of the ruminant, the absence of dung beetles and their restructuring effect is also notably lacking.

Rebuilding "humus factories"

> Primary climax ecosystems exhibit the highest net productivities of raw materials. They also have the highest rates of organic matter sequestration in the soil. A hasty conclusion from this observation would be to say that the more we return, the more we store. However, this overlooks the fact that these ecosystems primarily benefit from a true "factory" for producing humus through the presence of fungi. In our agricultural soils, which are degraded ecosystems, the challenge is first to restore these humus factories.

7.1.2 Ruminants break spore dormancy

Once excreted, dung is initially colonized by fungi belonging to the Zygomycetes family, then by Ascomycetes, which can utilize the remaining digestible cellulose. Next, Basidiomycetes settle in, equipped to break down lignin chains. With a pH of 6.5, an absence of fast carbohydrates, moisture for germination, and available bacterial nitrogen (not nitrates), mature dung provides an ideal culture medium for a wide variety of competitive Basidiomycetes.

Furthermore, the life cycle of many fungi includes passage through the digestive system of ruminants. Most species are present as spores deposited on plant leaves, well protected from UV radiation by brown pigments (often melanin). They are ingested in spore form by the herbivore. For some of these species, passage through the ruminant's digestive system is a means, if not the only means, to break spore dormancy. These spores are then ready to "germinate" as soon as the cow defecates. The loop is closed. Many would not germinate without this treatment. For example, *Pilaira anomala* only germinates 5% when spores are soaked in water at 37°C for 3 hours, whereas the germination rate rises to 100% when soaked for 3 hours in alkaline pancreatin, a medium resembling digestive juices.

7.2 Soils are not adapted to digest immature plants

The more you put in, the less there is... The implementation of green manures provides a good illustration of the paradox of organic matter levels in soils. In recent years, agroecology has highlighted the use of cover crops, with a neutral or even counterproductive effect on increasing organic matter levels in soils. "Overly green manures" indeed have the ability to counteract leaching of certain fertility elements. They can improve soil structure depending on species choices and implementation practices. However, it is a mistake to think that green manures significantly and sustainably increase organic matter levels for the ecological reasons already reviewed.

7.2.1 Restitution of immature materials

In nature, soils are not accustomed to digesting "lettuce"! Throughout millions of years of soil evolution, soils have never had to fulfill this role, or only in an incidental manner. Soils have thus not been able to develop efficient tools to digest these inputs. In "natural" ecosystems, the function of digesting immature plants is always entrusted to herbivores (such as insects and horses) and ruminants (as mentioned above). Under temperate climates, nature returns already digested organic matter to the soil from spring to autumn, and from autumn onwards, mature, senescent plant materials. These include leaves and dead branches. Apart from a few exceptions (such as aphid honeydew, inputs solubilized by leaching from tree leaves and trunks, etc.), this diet is consistently rich in plant fibers, cellulose, and especially lignin.

The paradox of excess nitrates and nitrogen starvation: too much and not enough ?

Consequence of climax lag: too much leachable nitrogen in autumn

Minerals that can leach come from fertilizers incompletely utilized by plants whose growth sometimes stops prematurely due to temperatures and summer drought. This leaching also results from the mineralization of humus, which continues after harvest (from July to October) in all soil layers.

Consequences of a lack of humifying fungi: not enough nitrogen at the surface

Fertilizers, all of which are salts (except urea), fungicides, green manures, tillage, and straw removal, are practices that favor bacteria. Bacteria are highly competitive: nitrogen scarcity will occur where there is fresh cellulose to degrade, thus in the surface horizon, at the level of a young plant's roots. This plant does not immediately have access to deep reserves..

Our agricultural practices often create a mismatch in space and time between nitrogen needs and availability!

7.2.2 Materials with low value for humus builders

By definition, "green manures" are plants that have not completed their life cycle. They are rich in sugars, cellulose, and soluble nitrogen. The carbon chains they deliver to the soil are relatively short. They have a fertilizing effect in the short term, but they do not constitute a primary source for the humification of organic matter by fungi and have a short turnover time (see figure). Instead, they provide a raw material that favors bacteria, which will mineralize the matter rather than humify it.

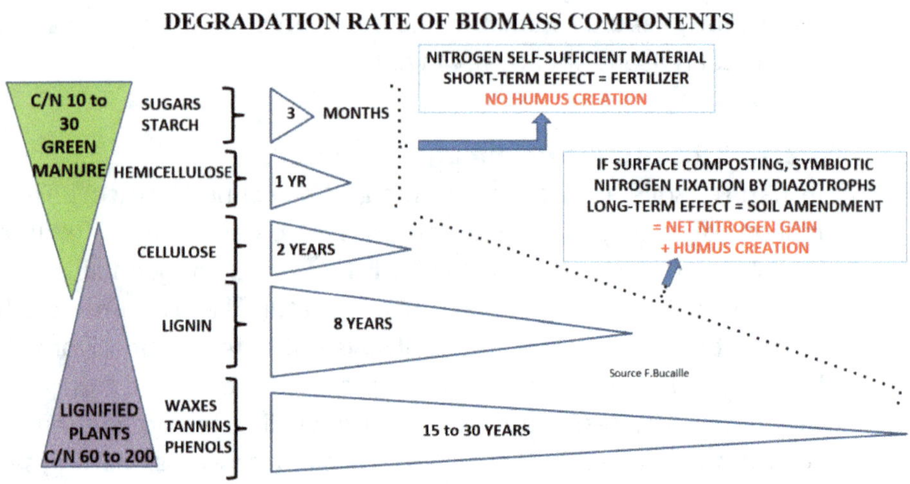

Figure 7.1 – Degradation rate of plant biomass components

The purpose of a plant is to reproduce itself

> The purpose of a plant is to ensure its descent and spread. For example, a plant in distress will always accelerate its flowering and seed production. Throughout its complete cycle, the plant incidentally provides other services such as combating leaching, erosion, or soil compaction. However, a plant primarily operates to serve its own nutrition, which is itself part of a comprehensive cycle extending to reproduction. It is this complete cycle that integrates the plant into its ecosystem by fulfilling all the functions it has performed for millions of years, including the ultimate "gift" of its lignin and celluloses after maturity and death. This is its contribution to the sustainability of the system, compensating for what it has taken during its active growth.

7.2.3 Late destruction at an immature stage: the penalizing scenario

When the cellulose content increases in plants, cellulolytic bacteria (actinobacteria, proteobacteria, including *Bacillus subtilis*) find themselves in an environment that is highly favorable to them, as long as the lignin is not at a sufficiently high level to prevent them from carrying out their work. They are indeed not equipped to attack lignin, which protects cellulose from the action of their exoenzymes. These same bacteria have relatively high nitrogen requirements (with a threshold of around 2% N). However, there is a stage in the plants' vegetative cycle where nitrogen levels decrease, while there is still plenty of unprotected cellulose available to them. It is at this point that cellulolytic bacteria may "seize" the available nitrogen in their environment to the detriment of a young crop. Paradoxically, the risk diminishes at more advanced stages of plant residue maturity, as the dominant actors, thanks to the competitive advantage provided by their lignolytic enzymes, are saprophytic fungi, which can tolerate much lower nitrogen levels (0.5 or even 0.2%). See Figure 7.2 below.

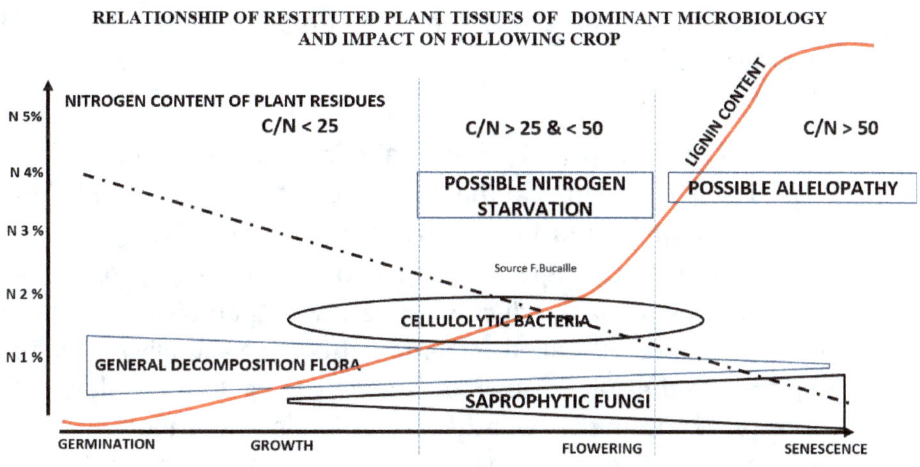

Figure 7.2 – Relationship between age of returned plant tissues – dominant microbiology and impact on the following crop

That's why there's no "nitrogen starvation" behind corn or under forest cover (and also the higher C/N ratio of forest humus: around 15). However, as the plant ages, there's often an accumulation of secondary metabolites for protection that have powerful allelopathic capabilities.

7.3 Mineralization of stable humus by plants

7.3.1 RPE: mineralizing effect of the rhizosphere

Plants have the ability to mineralize organic matter within their rhizosphere at different stages of their life. This mechanism of organic matter mineralization within the rhizosphere is now known as the Rhizosphere Priming Effect (RPE). Its effect ensures a higher level of nutrition for the plant, particularly beneficial from the first weeks of its development until grain filling. Modeling studies of the rhizosphere conducted on soybean, wheat, or sunflower models (Pausch, 2013) have highlighted this famous rhizosphere priming effect. The causes of this RPE are, however, less understood than the effect itself. Nevertheless, "clear evidence has been established of a link between RPE and enzymatic and microbial activation within the rhizosphere, with the trigger being root exudates" (Biao Zhu et al., 2014). It appears that all components of rhizosphere microbiology, especially bacteria and mycorrhizae, directly and indirectly participate in this phenomenon of accelerated mineralization.

7.3.2 Cover crops: an additional mineralizing effect

On average, it appears that the rhizosphere priming effect (RPE) increases the carbon mineralization (destruction) rate by 59% (meta-analysis of 31 experiments, Changfu Huo, Yiqi Luo, Weixin Cheng, 2017). Soils with finer textures tend to induce stronger RPEs than coarser-textured soils. Furthermore, the so-called activation effect can actually persist for a very long time in the soil throughout the active vegetative cycle of the plant, contrary to the initial hypotheses made since the identification of the phenomenon. The rhizosphere activation effect is crucial for regulating the biogeochemical cycles of soil carbon and nitrogen. We believe that implementing cover crops and maintaining permanent vegetative cover creates additional mineralization effects in the various agrocenoses where they are employed. The implementation of this approach has multiple benefits that are well-documented in terms of erosion and compaction. However, we believe that their mineralizing effect should at least be considered and deducted from the biomass produced, which has a very low isohumic coefficient (≤ 0.05). Their extreme practice involves transforming (through RPE) stable humus into short carbon chains (sugars, hemicellulose, celluloses) with limited half-lives through root exudates.

In Figure 7.3, we have indicated with a vertical line the latest stage of destruction of cover crops according to current practices. The RPE phenomenon appears to decrease from the grain filling stage, at which point all energy transfers, carbohydrate hydrates, and amino acids are directed towards filling reserve organs (grains), mainly through translocation.

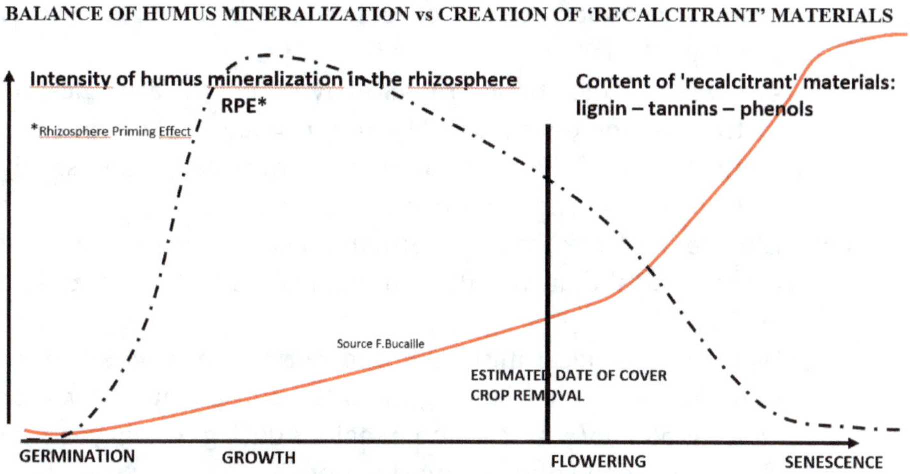

Figure 7.3 – Crossed curves between RPE mineralization and humus precursors (lignin)
Dashed line: Intensity of humus mineralization in the rhizosphere (rhizosphere priming effect (RPE). Solid line: Contents of "recalcitrant" materials (lignin - tannins - phenols)

As we can see above, the risk is to interrupt the plant's growth while it has triggered and achieved the essential mineralization of stable humus to ensure its growth, without having had the time to develop humus precursors, namely cellulose and especially lignin. When they are destroyed at too immature stages, their composition, rich in carbohydrates and nitrogen, will promote the development of bacterial families which are not the most efficient at storing carbon. Green fertilizers applied only at early stages, far from being able to sequester carbon in soils, may instead consume stable humus. If buried deeply, these green manures can also be a source of anoxia (due to a significant biological oxygen demand for their degradation) and soil deterioration.

The assumption is that by doing exactly the opposite of what caused the loss of fertility at the end of the 20th century, the result would inevitably be positive. The same assumption as in the 1950s: "the more we return to the soil, the more we'll gain".

In the 1950s, we had simply forgotten that, despite returning more straw and manure to the soil, deep ploughing would disrupt the strictly aerobic, lignin-hungry fungal flora of humification, thereby promoting mineralization processes.

The following questions arise today:

- Are we sure that returning insufficiently lignified green materials, rich in sugars, will not have the same depressing effect on this same strictly aerobic lignin-hungry humification fungal flora? That the permanently activated RPE (rhizosphere priming effect) will not take precedence by promoting mineralization processes and neutralizing the expected positive effects?
- Are we not exacerbating the already significant ecological mismatch by tropicalizing the seasonality of our practices (Permanent cover 12 months/year) and thus disturbing the microfauna and storage flora, with still poorly understood secondary effects?
- Is there not a risk that this permanent cover will serve as a refuge for pests (aphids) that will infect summer crops such as beets, lentils, and peas?
- Will this ecological mismatch and permanent vegetative cover not bring about another limiting factor by consuming water year-round, which is already scarce in some regions during certain years and could be much needed during critical periods?

Microbiology is difficult to encapsulate in equations and models. For reasons that may not be the same as those of the 1950s that we have just touched upon, it could still disappoint us if we are not careful.

7.3.3 Mycorrhizae and exsudates activate mineralization through RPE

We know that the plant fuels its growth by orchestrating its mineralizing rhizosphere through its root exudates feeding the rhizosphere, including the mycorrhizae. Some would like to present mycorrhizae as manufacturers of stable humus through the elaboration of their constituents: melanin, chitin, glomalin, lipids. They indeed receive simple sugars from the plant, which they transform into durable constitutive materials of soil organic matter, and they also store lipids received from their host. But directly and indirectly, they also actively participate in the mineralization of soil organic matter. They extend their hyphae into all compartments of the soil, solubilize organic phosphorus, participate in plant nutrition, which, boosted, increases root exudates to adapt the level of mineralization according to its needs and ensure maximum growth allowed by environmental conditions (heat, water, available nitrogen...). The correlation is so strong that it is believed this mechanism, which is the result of millions of years of coevolution between plants and microbes, accelerates the mineralization of organic matter with increasing CO_2 (Langley et al., 2009; Phillips et al., 2012), hence photosynthesis, hence yields. This mechanism can lead to a decrease in organic matter in soils through this RPE mechanism alone (Carney et al., 2007; Langley et al., 2009). Obviously, an exacerbated practice of permanent green manure can contribute to this counter-intuitive evolution. Favoring the best possible photosynthesis through regular foliar fertilization in the hope of storing carbon via transfers to mycorrhizae risks being thwarted by this feedback effect that neutralizes the expected positive effects. This fine plant-rhizosphere correlation, with immediate reactivity, is already widely proven by studies, one of which showed that under multi-species natural grassland, clear-cutting resulted in a 56% decrease in mineralization in the root environment within 24 hours (Shahzad et al., 2020). The reverse is also true. Increased photosynthesis levels go hand in hand with increased mineralization as well. **Generally, the balance of storage and mineralization fluxes in the rhizosphere tends to mutually balance each other out. To sequester carbon in soils, we will need to rely on the other biological lever formed by saprophytic fungi, active on dead plant residues.**

7.4 Conclusion: a tropicalization of practices

The practice of green manures - the principle of permanent soil cover - was first theorized and modeled in tropical environments (Brazil). However, soils in temperate regions do not function in the same way as tropical soils. We will see in the following section how the principle of an immature winter cover indeed represents a tropicalization of our practices and how this tropicalization generates a significant ecological mismatch for the soil system.

8 • OTHER PATHWAYS AND CONCLUSION

In recent years, we've seen a growing trend in the field towards practices that enrich soils with organic matter from exogenous sources. The use of forest ecosystems and hedgerow wood in the form of RCW (Ramial chipped wood) has produced good agronomic results, as has mulching with cereal straw. The good results obtained with these practices are further proof of the value of providing the soil with a ration made up of mature, cellulosic and woody plants. On the other hand, we note that they involve a transfer of fertility that is impossible to generalize.

We're seeing a revival in biological activity, with earthworm populations soaring. All this is accompanied by plants that are less susceptible to disease and drought. It would even seem that this famous "nitrogen starvation" is not appearing, as should theoretically be the case. Approaches of this type in market gardening are in the process of demonstrating this in real-life situations. However, the quantities involved are often enormous, and involve a transfer of fertility. In this case, we often speak of recycling waste with no added value. From an economic point of view (the price of straw, tree bark or branches is relatively low) this is true, but from a biological point of view these woody materials are the most valuable for the soil, as we have just seen. That's why these techniques work so well.

8.1 RCW[10] and mulching: mature materials

8.1.1 Practices with little or no general applicability

RCW (ramial chipped wood) is regularly used, with inputs of between 100 and 800 tonnes per hectare per year! But RCW resources are limited. Belgian foresters, for example, have estimated the production potential under sustainable management at 137,000 t/year. At the doses often used, this corresponds to around 1,000 ha. So this is not a practice that can be widely extended. It can only be used to enrich a small amount of land dedicated to specialized, high value-added production. The same applies to certain practices advocating exogenous straw inputs of 20 to 100 tonnes of straw/ha/year. This presupposes "stripping" 5 to 25

[10] RCW is a forest product composed of woodchips made solely from small to medium-sized branches. It is used in agriculture for mulching and soil enrichment.

hectares of cereals (or even more) of straw to fertilize one hectare. This is far from a virtuous circular economy system. Conflicts of use could very quickly arise. In fact, more and more farmers are recognizing the importance of their straw for their land, and are refusing to export it (except in the context of a straw/manure exchange).

Concentrating organic by-products from outside areas on a small surface area is not a matter of self-sufficiency, but of fertility transfer. This is made possible by the added value of the soil products sold, and sometimes even by the valorization within the framework of a particular business model where these experimental platforms serve as an illustrative basis for training services, conferences, influencers, or publications.

8.1.2 Forests operate with only 3 t/ha of dry matter

The annual productivity of a temperate forest is estimated at between 2 and 4 t/ha of dry litter matter, of which 50-80% is leaf litter, 3-30% is twig and bark litter and 1-30% is miscellaneous litter (Mangenot, 1980). Contributions from trunks are not taken into account. In principle, they are negligible in a managed forest. Root inputs are more difficult to assess. Herbaceous inputs and inputs from mosses are of little importance in terms of quantity. Assuming that these inputs are negligible, we can see that the forest is capable of self-maintaining its fertility with 50 to 200 times less input than the recommended RCW input! In addition, cultivated systems are capable of providing the soil with restitutions equivalent to forest inputs (3 to 4 t/ha of dry matter).

8.1.3 Soils don't need as much

There are solutions that avoid robbing Peter to pay Paul. They rely on the establishment of coherent, comprehensive systems where the efficiency of each of the main soil functions (humification, mineralization) is optimized, and major obstacles are removed (allelopathy, mineral deficiencies, compaction, hydromorphy...). Christine Jones has demonstrated that agricultural soils can not only be carbon-neutral but can also enter into a dynamic of fertility restoration, without external inputs. The problem of declining organic matter in soils is not merely an accounting issue of inputs and outputs. It is a problem that finds its answers by taking into account soil ecology. In this case, fungi (mycorrhizae) are essential. They are the ones who build humus: the cell walls they generate have half-lives of thirty to forty years. They store carbon durably in the soil. Their mycelium is described as the "Internet" of Nature (cf. Paul Stamets). They ensure communication between plants and the transfer of nutrients from one plant to another. They buffer the physical and chemical excesses expressed abruptly by parent rocks (acidity, alkalinity, toxic metals). In a field, they thus reduce spatial heterogeneities and ensure balanced crop nutrition. But these populations are also the most penalized in agricultural soils. Plowing/soil cultivation, fungicides, soluble fertilizers... All these agricultural practices inhibit them. It is their inhibition that explains most of the impoverishment of agricultural soils, especially in stable organic matter. Before adding more and more organic materials to the soil, let us ask ourselves whether the soil is capable of transforming them into stable humus. Then, let us adapt our practices to favor this particular food chain by preserving mycorrhizae as much as possible.

8.2 Saprophytes are the key to humus

We have seen that the rate of carbon storage in tissues varies greatly depending on the actors involved. Regarding bacteria, 15% of the carbon is preserved, whereas fungi store 50% of the carbon in their walls, and moreover in stable organic compounds. The challenge lies here, and this is where all the variability in the expression of iso-humic coefficients used in balances resides. 0.05 or 0.15 or 0.20% efficiency? It all depends on who is responsible for the evolution of organic matter. What degree of maturity of the plants triggers the intervention of which actor?

We have seen that there are three major functional families in soils:

1. **Symbiotic organisms** (such as Azotobacter and mycorrhizae) which directly contribute to crop yields by collaborating with them.
2. **Saprophytic organisms** (such as cellulolytic bacteria and basidiomycetes) which primarily contribute to the efficiency of carbon storage by participating in the decomposition of plant residues.
3. **Parasites** which translate any nutritional imbalance into destruction of living plants.

All actors are important: **symbiotic organisms** are essential for crop productivity and the economy of the system, but we have seen that the balance of their activity in the carbon cycle is neutral. Their increased activity results in increased mineralization of soil humus.

Saprophytic organisms are those where we have a huge potential for improvement (decomposing bacteria, especially ascomycetes and basidiomycetes). They hold the lever for the transformation of dead plants and animals and influence the rate of transformation of fresh organic matter into humus.

In the era of climate change, the impact of cultivated ecosystems and soils in the carbon cycle is a source of both anxiety and hope. It's a source of anxiety because soils store two-thirds of the Earth's carbon, and ongoing processes in soils lead to a weakening of organic matter levels. On the other hand, it's a source of hope because research shows that soils' capacity to capture organic matter alone could help capture the excess CO_2 generated by human activities. The processes of plant photosynthesis and soil restitutions have the ability to economically remove more carbon dioxide from the atmosphere than any other technology. It's from this strong idea that programs for CO_2 storage in soils are being implemented (such as the "4 per 1000" initiative initiated by France at COP 21, for example).

With a favorable evolution of organic matter, as we are seeing with the restorative practices we are implementing, these objectives seem entirely attainable to us.

Healthy soil ? or productive soil ? Both!

For a soil to be consistently productive, it must also be in good health. A healthy soil is one in which the biological processes that have contributed to its formation are still active, respected, and restored if necessary. This includes the flow of water, the cycle of organic matter, mineral balance, and the nitrogen cycle. It corresponds to situations where plants delve deeply into the subsoil, ideally reaching the bedrock, and where crop rotations and their management mimic natural climax ecosystems as closely as possible.

C

The climax: a key to intensifying a sustainable agroecosystem

Agricultural soils have been developed over the past 10,000 years on original ecosystems known as climax in ecology. The climax represents the ultimate evolutionary stage of balance between climate, soil, and the species living there. Biodiversity, productivity, and functionalities are at their peak. It is extremely beneficial for farmers to refer to the climax of the region where their land is located to assess its potential and identify everything that can be restored in terms of major natural functions that have been diminished or lost. This is where the new sustainable reservoirs of productivity lie.

9 • PRODUCTIVITY THROUGH APPLIED ECOLOGY

9.1 New foundations for soil ecology

The soils we inherit are the result of these balances, which took tens of thousands of years to establish. Our practices have more or less severely demoted them to earlier, transitory evolutionary states. However, their natural dynamic is always to restore themselves. This occurs with the help of pioneer species, some of which are easily recognizable to us by the disadvantages they impose; these include weeds, pests, and parasites.

9.1.1 Each in its ecological niche

■ Gause's law

Ecology and observation of nature teach us that within these stable ecosystems, each species occupies a very narrow ecological niche (a place within the ecosystem including the environment and interactions including feeding habits). Indeed, according to Gause's law (1934), two species with similar ecological niches cannot coexist simultaneously and durably. In the long term, the processes of selection would inevitably lead to the disappearance of the less adapted species to the detriment of the other. It follows that within an ecosystem, **each species has a unique dietary regime with possible but limited overlaps.** Furthermore, in these stable ecosystems, the absence of one of the links in the chain is likely to disrupt the entire system.

■ Functional roles for each species

Predation, primary production, secondary production... In the grand theater of ecosystems, the distribution of roles has been fine-tuned over thousands of years. Each species fulfills a specific functionality within the ecosystem. This role takes different forms from one species to another, but the result is always to counteract any excesses that may occur. The cheetah seizes the sick gazelle before it contaminates the herd, the vulture cleans up carcasses, ruminants graze and return to the soil feces rich in lignin and other indigestibles, protozoa devour bacteria, certain fungi consume nematodes... Each species needs the others for its own survival. The prey needs the predator as much as the predator needs the prey. These self-regulating functions of the ecosystem on itself are ubiquitous.

The expression of biodiversity sometimes gives the impression of a chaotic and exuberant proliferation of disorganized species. This is not the case. In nature, everything is extremely well organized. **Biodiversity acts as an effective "police force"** that allows everything that strengthens the great balances to flourish and represses anything that could oppose them.

■ **Parasitic infections are the red flags of our agroecosystems**

Within an ecosystem, there are no "good" and "bad" entities. All forms of life have their usefulness, provided that each of them remains within its field of activity in a balanced, stable, and sustainable system. In soils, bacteria, worms, insects, crustaceans, and fungi all belong to a food chain and are terribly interdependent. Parasitic infections thus make us aware that something has not worked in the plant's nutrition. Their appearances often contain part of the solution for the cultivated system itself. If we see a pest arriving, it means that we have opened its ecological niche. Understanding the process is essential to then hope to be able to close it again. In 80% of situations, this is related to the soil system.

9.1.2 Ecosystems that have regressed to an unstable pre-climax state

■ **The "climactic" forest model is never sick**

During evolution, and before significant human modifications, ecosystems evolved to reach points of stability. This ecosystem, having achieved a certain fragile, evolving balance between the climate, soil, and species, is termed a climax by ecologists. Within the plains of temperate and humid climates, the climax most often corresponds to forests. As we have seen, in these climactic ecosystems that have found their balance over thousands of years, everything contributes to the equilibrium of the environment. These systems are never sick. What we call parasitism in them is part of extremely fleeting, circumscribed, and temporary ecosystem regulation mechanisms.

The boxwood moth, processionary caterpillars, cryptogamic diseases, etc. Our current temperate forests are not spared from the resurgence of parasitic diseases, which contradicts the often accepted idea that "forests are never sick." However, it is not so much the forest that is never sick, but rather the climax, which is the point of culmination and stable balance between species and their environment. Cultivated forests represent an intermediate situation, which, when not concerned with the nature of endemic species, can also have disastrous effects on the environment and soil evolution. Soil-vegetation balances have been built over millennia. For example, conifers are adapted to high latitudes or altitudes. Acidification, depletion of associated fauna, microfauna, and microbiology. Planting conifers where deciduous trees once grew is often detrimental to the ecosystem. In most regions of the world, the forest ecosystem is no longer a climax ecosystem, and large ecological niches have been opened to parasitic diseases. Forests are often heavily influenced by human activities. It seems that the only "surviving" primary forests in Europe today exist in Finland, Poland, Ukraine, but also in some remnants in most European countries, including Mediterranean countries.

The planted or selected tree species are not always the most suitable for the ecology of the site. The self-regulating functions among species are less effective there. This is especially true when new exogenous species have been introduced. Due to human activity, forests are also the scene of significant fertility export flows. Thus, even the forest ecosystem can be ecologically out of sync with its soil, climate, and entire biocenosis. This mismatch results in forest ecosystems having their ecological functionalities weakened. Once again, it seems to us that it is the responsibility of humans to intervene in the major ecological functions of the forest that they have compromised. Forests also deserve to be restored, even if the time scale is much longer than for annual crops.

■ Absence of the three “P”s (perennial, profound, and permanent)

We have seen that most arable lands worldwide have been reclaimed from former forest ecosystems. Soils in cultivated ecosystems thus lose the benefit of the presence of the climax biocenosis and the major ecological balances that counteract excesses expressed in the form of parasitic diseases or stress. In addition to this, soils lose the benefit of having trees with perennial, permanent, and profound (deep) root systems (the 3 Ps). The recycling of deep minerals and their recirculation to the surface is much less effective without trees. The man who cultivates thus removes from the soil a very significant part of its capacity to regenerate itself and to resist surface acidification (linked in particular to the departure of the bases at depth). The absence of a canopy or organic plant debris no longer effectively protects the soil surface from the "splash" effect caused by raindrops. The establishment of mycorrhizae, water functioning, etc., are also weakened.

■ Simplified environments sensitive to excess

The establishment of agriculture also results in a significant simplification of the environment. From polyculture/breeding, there has been a generalization towards specialized breeding and polyculture with rotation, and finally monoculture. This simplification is reflected in the soil microbiome. In addition to the absence of deep roots, agricultural practices regularly have the effect of reducing the diversity of biological functions in the soil. Buffers and barriers disappear and the environment is more sensitive to excesses in all its forms (drought, pH, hydromorphy, salinity, etc.). These conditions favor transitional pioneer species, often dominated by the most robust.

We favor pioneer and extremophile species

> The consequence of anthropogenic disturbances on ecosystems is to systematically favor life forms that appreciate extreme conditions (extremophiles) and are more autonomous (autotrophs), very often pioneers (anaerobes, archaea, invasive weeds, etc.). These life forms respond to the new conditions of a retrograded soil. They fulfill the ecological function of restoring the initial state by constructing the first stages and preparing the following ones. If we let it happen, it would take around 300 years to rebuild a "secondary" forest that would resemble the primary forest. Their excessively large presence is an indicator of impoverishment and fatigue of the natural functions of the soil. Furthermore, in the case of annual crops, each year the ecosystem must start from scratch, at least on the surface. This implies that huge ecological niches are opened up every year to new "intruders". However, a large part of the functionality of the natural ecosystem can largely be maintained or restored. So why should we be condemned to living with parasitic diseases? Why should the efficiency of plant feeding necessarily be lost? Why should soils be condemned to erosion and loss of structure? All this is not inevitable. These phenomena are so common that today we consider them normal. In reality there is nothing normal about them, they are warning messages. We should never shy away from looking for the root causes.

9.1.3 Intervention is essential

■ Beware of the temptation of "mimetic" agrosystems

Biomimicry is an applied science aimed at innovating by drawing inspiration from natural biological mechanisms. The zipper, Velcro, the nose of the Japanese Shinkansen, the needles of the latest generation syringes, honeycomb materials, antibiotics... All these innovations might not have come to fruition without the technological inspiration found in nature. However, dressing up as a bear has never improved strength. Dressing up as a kangaroo to jump higher has never worked. So why would dressing up as a tropical forest (see next chapter) when you are a European ecosystem have a better chance of working?

In the absence of a deep soil cation recycling process (provided by large trees), it may make sense to restore calcium fertility by liming, or to consider agronomic plowing (see our definition below). We can see that **a soil function can be restored by very different means from those used in the original agroecosystem. This is the meeting point of two techniques: biomimicry and functional analysis.**

Trying to shoehorn cultivated ecosystems into the mold of the "natural" ecosystem is an appealing idea in theory, but it quickly comes up against field observations and common sense. Soil structure, absence of surface acidification, absence of diseases, etc. Nature, the climax, is indeed our model, our ideal, our source of inspiration. Our goal is to recover equivalent functionalities in the cultivated ecosystem. But this doesn't necessarily involve mimicking nature. Otherwise, veering off track is not far off, as seen with the increasingly frequent emergence of no-till hardpans in soils (see Part D, Section 13.1). Synchronization of plant growth with precipitation, residue management... In the next chapter, we will provide some pointers for limiting the effects of the ecological mismatch that we impose on ecosystems.

■ Let's not confuse the ends with the means

It is necessary to assume that measured and calculated interventionism be considered, to understand the mechanisms that oppose the restoration of expected functionalities, and to adapt practices accordingly. Techniques known as conservation and regenerative agriculture are excellent, provided that the objectives pursued are fully understood. It's not always enough to have a direct seeding drill and cover crops to compensate for the natural functions weakened by the cultivation of the ecosystem. Let's not confuse the ends with the means. Direct seeding or conservation tillage systems are means among others, not the end goal. They deserve to be accompanied, evaluated by facts and observations, and used judiciously.

When failure to act is a crime of non-assistance to an endangered ecosystem: the Loire riparian woodland

Certain alluvial sites, often classified as Natura 2000 (a geographical mapping of sensitive areas in France which are protected by regulations and limited authorized operations), are subject to agri-environmental and climatic measures (called MAEC regulations in France) which prohibit the use of fertilizers and soil improvers of any kind, for "ecological" reasons, to maintain the natural character of the area and its biodiversity. One of these MAECs concerns the Loire Valley, a long river that rises in the Ardèche on soils of volcanic origin. As a result, the river rolls over acidic, non-calcareous sands along its entire course. Long before it reaches the ocean, it crosses regions with limestone subsoil at Nevers, Tours and Angers. The vegetation formerly naturally present is a riparian forest of willows, ash trees, trees with deep roots. These species are capable of absorbing calcium from the limestone bedrock and then transporting it into their trunk, branches, and leaves, which fall to the ground. This is the natural and virtuous process of recycling bases on the surface for this ecosystem. Today, these mostly meadow soils (as they are floodable), no longer benefiting from the presence of trees, are weakened in terms of the recycling of deep minerals. Limestone amendments would restore the proper functioning of these soils, which is prohibited by the MAEC. Without human intervention of this nature, these meadow spaces are

> nevertheless doomed to see their soils sink into a process of surface acidification and to see the flora with superficial rooting deteriorate. The intentions of the MAEC are commendable, but the non-interventionism advocated here amounts to failure to assist a perilous ecosystem. The ancient deforestation of these places requires humans to replace the mechanisms they have destroyed. This is the responsibility of the domesticator and the clear-cutter, in addition to being an ecological imperative.

9.2 Managing the agroecosystem's climactic lag on its soil

The agricultural ecosystem will always be out of step with the climactic ecosystem. Managing and mitigating the effects of this misalignment based on scientific ecological knowledge constitutes a new, inexhaustible, and sustainable reservoir of productivity. This must be done by taking into account the facts and the "natural" model as a whole and translating it into:

- Identifying sequences and processes that are atrophied.
- The intervention of the farmer to restore them.

9.2.1 In temperate regions, winter vegetative pause is necessary

■ Tropical and temperate ecosystems: very different cycles

The scientific figures and popularizers currently influencing agricultural practices in Europe and elsewhere have drawn their inspiration from the conservation or regenerative agriculture models of equatorial and tropical regions, primarily in Brazil. They have established some postulates, which are:

- Permanent soil cover;
- No-till and direct seeding;
- Crop diversity and rotation.

In the background, and complementarily, the use of localized fertilization is advocated for. Other schools of thought aim to promote root exudates, the development of mycorrhizal fungi if necessary by resorting to repeated foliar fertilizations. We have already expressed our reservations about certain aspects of these practices, which aim to improve soil structure and, above all, to store carbon in agricultural soils. These approaches offer major improvements over a deep-tillage system, with plant debris buried and the soil left bare for six months of the year. However, it seems to us that adjustments are necessary for the targeted objectives to be achieved. Before considering these adjustments, let's first take a look at our climatic regime in temperate zones.

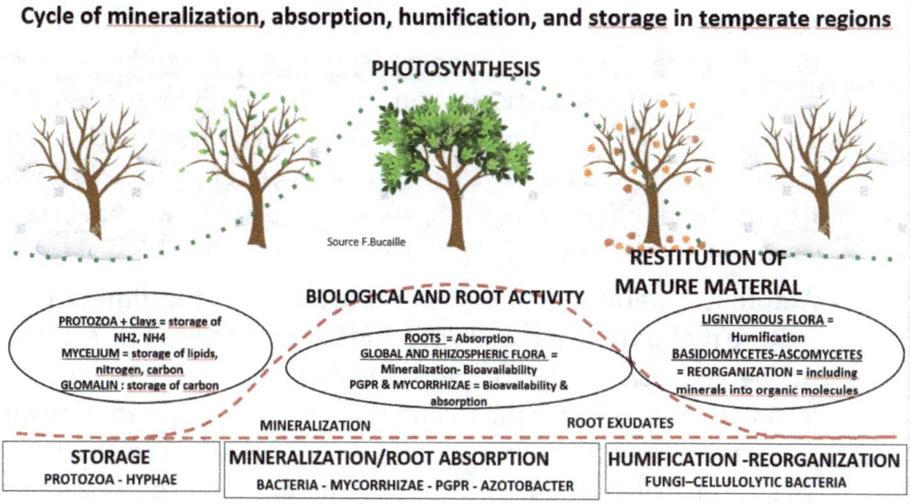

Figure 9.1 – Cycle of mineralization, absorption, humification and storage in temperate climaxes (PGPR [*Plant Growth Promoting Rhizobacteria*]: bacteria promoting assimilation and therefore mineralization) (trees: MarBom/shutterstock.com)

Permanent ground cover with green, active vegetation does not correspond at all to the climax of temperate zones. This is the model for tropical regimes.

In tropical and equatorial zones, the natural model is to store everything that can be washed away in the form of plants, animals, microfauna and microbiology. The kaolinite-type clays in these very old soils with low CEC (40 to 100 meq/kg), low retention capacity, exacerbated mineralization 12 months out of 12, abundant rainfall (up to 3,000 mm/year) mean that **the only possible storage for anything soluble is the living itself.** The simultaneous processes of mineralization and humification are permanent. Permanent cover is an obligation. In Brazil, this allows up to three harvests a year: a wheat crop, a "large" corn crop and a "small" corn crop.

■ The temperate cycle alternates humus storage and mineralization with a winter break

In our climates, with their highly differentiated seasons, other processes are at work, quite different from the tropical cycles (see Figure 9.1). Here's how the seasons work in temperate soils:

- **Spring and summer**: a **period of intense mineralization** and mobilization of all commensal and mutual mechanisms around plants. In summer, the soil's mineral, humic and water reserves are released to fuel plant growth. This is when the soil is covered and protected from the summer sun's rays. There is no bare soil. The epigeous life forms are in the shade of the foliage. Worms, carabids, springtails, microfauna, microflora, yeasts, algae (cyanobacteria) can thrive. Everything is geared towards chlorophyll activity, encouraged by mycorrhizae and rhizospheric microflora. Mineralization triggered by rising temperatures and the RPE (rhizosphere priming effect) trigger soil → plant flows that exceed plant → rhizosphere flows.

- **Autumn:** a **period of humification**. Autumn is the time for returning mature materials rich in cellulose and lignin to the soil. The microbiology of decomposition is at work: cellulolytic bacteria, ascomycetes, basidiomycetes. This is the moment when the "loan" made by the soil to the plant during the summer is repaid by the return of woody material. This is the crucial moment when the humic balance must be restored. It's the moment when the organic matter level decreases or increases. **This is where the game is played:** if active vegetation (such as green manure) is still in place, mineralization through RPE will continue to be intense, and depletion will persist. This is also where we reap the benefits of the year's crop management:

  If no fungicides are applied at the end of the cycle, if mineral balances are correct (P, S, N, see chapter 10), if straw is not exported, if mature crop residues are returned, then fungi specialized in decomposing these materials (ascomycetes and basidiomycetes) will repay the "loan" with interest. Their carbon recycling performance from plant residues is 50%. **The humic balance will be positive.**

  On the contrary, if the competitive advantage is given to decomposition bacteria due to one or more of the following causes: straw with fungicides, glyphosate, excess nitrogen and/or phosphorus, restitution of green materials rich in simple sugars, waterlogging, and hydromorphy, then the "overdraft" will increase. The performance of bacteria in recycling, reorganizing carbon is only 15%. **The humic balance will be negative.**

- **Winter:** a **storage period**. Winter marks the end of the decomposition process. It depends on the ambient temperature and the nature of the materials to be decomposed. The ground temperature is maintained thanks to the shelter of trees against the wind, and sunlight reaches the ground after the leaves fall. These soils do not support active vegetation, but they are not bare either: there is abundant litter on the ground. It's also a time for rest and storage. Many of our temperate brown soils have high CEC (100 to 350 meq/kg), therefore with much greater cation storage capacities than tropical ferralitic soils (40 to 100 meq/kg). Storage mechanisms also depend heavily on microflora and microfauna. In winter, fungal mycelium itself becomes storage centers: their walls composed of glomalin, chitin, and melanin preserve them from rapid degradation (with half-lives that can reach ten to forty years). The nitrogenous materials and lipids contained in these hyphae are thus protected from leaching. Hence the importance of promoting high fungal populations. Furthermore, the significant mortality of bacteria due to the cessation of vegetation (cessation of the RPE effect) and the decrease in temperatures, as well as their intense consumption by protozoa, generate the production of ammoniacal nitrogen and amines (NH_2 and NH_4). When these forms of nitrogen become free, they are stored on clays, as other cations would be. These forms of nitrogen are not leachable. As long as the temperature remains low, and as long as a plant does not activate mineralization, they will remain stored. Obviously, these storage processes do not apply to residual nitrogen due to poorly calibrated nitrogen fertilization, nor to nitrates resulting from mineralization the previous summer and not used due to lack of vegetation, since nitrates are not fixed by the CEC. When microbiology is active and the bacteria-fungi ecosystem is operational, this resource-saving mechanism, which dates back hundreds of thousands of years, comes into full play. This is also the water storage period. In non-irrigated situations, and/or in years with low rainfall, this water must not be consumed by a winter and early spring cover crop, which is a real "water pump". This water can be sorely lacking for spring/summer production.

Taking into account the ecological role of the winter break

The seasonal mechanism specific to temperate regions - with a vegetative pause in winter - should be taken into account and its functions "mimicked" in soil revitalization programs. Models should take into account the importance of this much underestimated decomposition fungal flora, which is at the heart of carbon storage. All amines and ammonia are perfectly stored in the soil without loss. We'll also take into account the impact of the vegetative stage of the material returned to the soil, and the seasons when plant cover is most effective. Here are a few examples of how revitalizing agriculture can be applied in different agrosystems, inspired by our own climax.

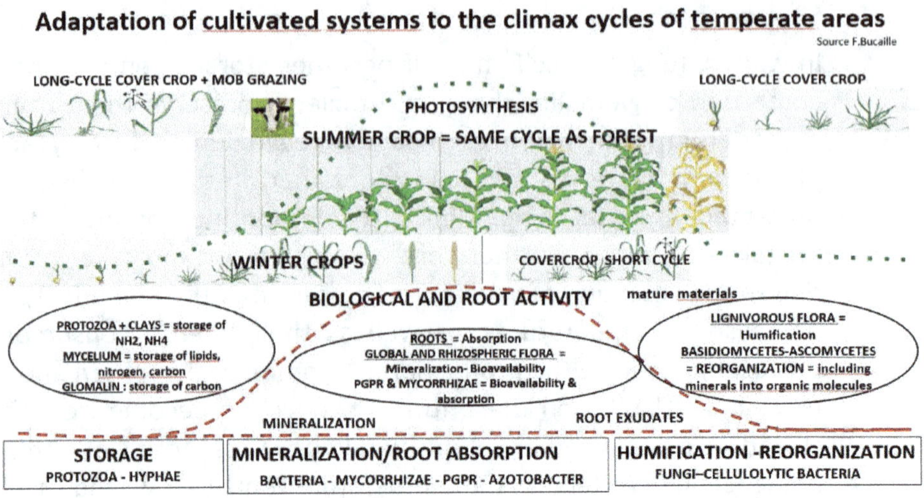

Figure 9.2 – Adapting farming systems to temperate zone climax cycles

9.2.2 Tillage is a historic and modern key to good ecological management

The practice of tillage is often criticized nowadays. However, it remains one of the historical keys that humans have found to restore certain fragile functionalities of cultivated soils. Indeed, through agronomic plowing, our ancestors understood that they could bring up elements to the surface that would otherwise leach away, thus managing the absence of recycling of deep minerals by large trees. There's a growing notion in no-till farming that regular liming is beneficial to neutralize surface acidity. Reintroducing agronomic plowing in certain situations, breaking up soil compaction when necessary, or leveling out ruts are possible responses.

■ No-till is sometimes excessive for the ecosystem

It is a counter-current thought at the moment, but not tilling the soil is sometimes an ecological mistake, because the activities of moles, voles, foxes, and dung beetles no longer create macroporosity, and the strong and permanent roots of trees no longer prevent the collapse of fragile soils (sandy loam).

Agriculture is evolving and taking a turn, which is fortunate. However, there are some pitfalls to avoid, it seems to us. There was a period when everyone was trying to plow as deeply as possible with 16-inch, then 18-inch furrows, with the practice being glorified through plowing contests. The criteria for crowning the champion: well-thrown furrows, deep and even plowing, and especially not a single straw on the surface. This type of plowing removed the benefits of the exercise. Today, we are witnessing another race for records: who can produce the most cover crop tons per hectare, who can produce the most "lettuce" per unit of surface, who has the most exotic plant in their mixtures. A pendulum effect from one excess to another. Agroecosystems need to benefit from the "mediety" described at the beginning of this book. Today, we meet farmers suffering from the "syndrome of the attentive pupil" described by J-P. Sartre in *Being and Nothingness*: "so focused on the objective of being attentive to the teacher's words that he no longer retains the content, as his entire being and all his energy are crystallized towards the goal of being attentive." Similarly, some are so busy having greenery all year round that they forget what their soil is whispering in their ears: plow pans (described in § 13.2.2), summer water deficit... We believe that this pendulum movement, from one extreme to the other, stems from some major misunderstandings in the interpretation of facts, causes, and effects, particularly concerning soil tillage."

■ The virtuous plowing of our forefathers

□ Our plowing ancestors did not destroy the soils

For centuries, man has plowed. For a very long time, this plowing was - let's not be fooled by the words - no more than "superficial digging". We can't pretend that these long practices have left 20th century farmers with humus-poor soils, devoid of earthworms and with a collapsed biodiversity. Everyone agrees that it was in the mid-20th century, in the 1950s, that fertility indicators began to falter. Agronomy is about taking a long-term view, and for reasons of scientific rigor, this should encourage us to look beyond the randomized trials of the last forty years. This is why taking a look at the past and reintegrating this enormous mass of experience into the construction of the third way in agriculture can save time and avoid mistakes.

□ Superficial plowing

Following the agronomic reflections of Olivier de Serres (*Théâtre d'agriculture et Mesnage des champs* [*The Theater of Agriculture and Field Management*], 1600), Diderot described good agronomic practices for plowing in much the same terms in his encyclopedia.[11].

■ Deep plowing by "rogue farmers"

"A four-inch plowing is enough to turn over the turf, expose the grass

[11] "Labour", *L'Encyclopédie*, first edition, 1751, p. 146-147.

roots to the air so they dry out and the plant dies."

Four inches... The plowing mentioned here is ten centimeters! A light tilling in today's terms. This is nothing like the caricatures of plowing that have been popularized lately. And here is what Diderot adds when talking about deep plowing, beyond the vegetative layer:

"The depth of plowing must be proportionate to that of the humus or topsoil, to the needs of the seed to be sown, and to the circumstances that determine the plowing, primarily the depth of the humus. There is a considerable number of lands suitable for wheat production, although they are only six to seven inches deep. **If you dig deeper, you bring to the surface a kind of clay** which, without being infertile, makes your land unsuitable for wheat production. .../... The first lifted layer carries with it part of the second; and it is then said that the soil is unbound. **Roguish farmers who are forced to leave their farm unbind the soil of their lands that can be unbound during the last two years of their lease. By doing so, they harvest more small grains and at the same time harm the person who is to replace them."**

Isn't that exactly what we've been doing? Haven't we been playing the "rogue farmer" for fifty years? Isn't it curious that, although the practice of plowing in the eighteenth century and the twenty-first century are so different, it's plowing as a whole that has been called into question? Imagine that we've gone from plowing ten centimeters (4 inches), maximum fifteen centimeters (6 inches), done in conditions of ideal humidity and never subjected to the destructive force of our mechanical means, to plowing 35, 40, 45 centimeters sometimes in the worst conditions, like sometimes behind beet (in Diderot's time beet didn't exist. It was Benjamin Delessert, an industrialist, who launched this crop at Napoleon's request to ensure the supply of sugar to France, which was subject to the English naval blockade).

■ Virtuous management of cover crops through grazing

Diderot: "The English have given us the most encouraging example in this regard: since artificial pastures have multiplied their herds and fertilizers, their harvests have increased to a point that would be doubted if one could refuse to believe the witnesses who attest to it. .../... By allowing the herds to continuously graze on the grass that regrows there, in some provinces, intelligent farmers borrow sheep from those who are not. They purchase the right to have them live for a certain time on their lands; and the abundance of crops is always the result of this arrangement."

The best use of green manure had already been discovered by the agronomists of the time: non-lignified materials are not returned directly to the soil. The “intelligent farmers” knew this. They were completely unaware of the existence of bacteria-fungi ratios or the action of digestion on plants and microbiology as we have described it, but their keen observation enriched by generations of farmers had led them to this conclusion. Having played the **“rogue farmers”** with anti-agronomic plowing, are we not now moving on to Act II and playing the scene of the **“unintelligent farmers”** with our current management of crop successions and green manure?"

10 • CONSIDERATION OF CLIMAX LAG FOR THE MAIN TYPES OF CROPS

We have seen how important it is to take into account the original ecosystem, the climax, in establishing our agricultural practices. Here we provide some ideas for reflection and action to better manage the ecological mismatches induced by different types of agricultural cover crops. We will therefore address the following agroecosystems (in temperate environments):

- Cereal crops with straw;
- Summer crops (corn, sorghum, sunflower, soybean, millet...);
- Green manures;
- Pasture land.

10.1 Cereal crops: mastering a climactic lag in temperate climates

The majority of plants cultivated under temperate climates with high rainfall (especially cereal crops with straw) are plants that utilize 500 mm of water per year. Additionally, the seasonality of these crops is asynchronous with these precipitation patterns. These mismatches require special management by farmers.

10.1.1 Wheat is an exotic Mediterranean plant

Agriculture, along with wheat cultivation, emerged approximately 10,000 years ago. During this time, humans began cultivating the first cereals found in the wild. Through the selection of the most beautiful ears, two types of "wheat" emerged: einkorn (*Triticum monococcum*) and emmer (*triticum dicocooides*) both originating from an area that currently includes Turkey, Iran, Israel, Jordan, and Syria. Then, a new spontaneous crossbreeding occurred between emmer and a wild grass species (*Aegilops squarrosa*) also native to the same fertile crescent. Its genome combined with that of wild emmer, giving rise to a new species: *Triticum aestivum* which would later become winter wheat. Concurrently, emmer wheat gave rise to durum wheat. There were numerous exchanges of wheat among Egyptians, Greeks, and later Romans around the Mediterranean basin. With the assistance of Christianization and Roman colonization, wheat cultivation as a staple food spread northward, replacing hazelnuts, acorns, and chestnuts, which were less productive but better adapted to the soil and climate than wheat.

How the "bucaille" arrived too late and missed its entrance

> Buckwheat ("bucaille" in Old French) is a plant with a summer cycle that is more adapted to temperate climates than wheat. Unfortunately, originating from China and Nepal, it only arrived in Europe in the 14th century. The cultural value of wheat was already well established. Wheat, originating from Mesopotamia and Palestine, has long been laden with strong symbolism; that of unleavened bread, Sacramental bread, and the French baguette. Buckwheat, known as "bucaille" in some regions, started with a certain handicap and some major flaws. Devoid of gluten, it cannot be used for baking bread (it can only make flat pancakes called "galettes"); with an indeterminate growth habit, it continues to grow until frosts, resulting in seeds of varying maturity. Despite its numerous other qualities such as adaptation to temperate climates, frugality in nutrients (especially nitrogen), resistance to diseases and insects, and competitiveness against weeds, this crop arrived too late in the West and failed to establish itself. Selection did not show interest in buckwheat and did not improve its qualities or reduce its weaknesses.

10.1.2 Lower, asynchronous water requirements

Despite its symbolic aura in continental Europe, wheat remains an "exotic" plant in northern Europe. Its origins are Mediterranean, and its water requirements (between 300 and 500 mm per year) correspond to savanna or steppe zones. Establishing small grain cereals in these areas, such as the Great Plains of North America or Kazakhstan, is ecologically relevant. However, doing so in our climatic zones leads to significant ecological mismatches. These crops do not use enough water, and especially not at the same times as the naturally associated biocenoses. Furthermore, the period of tissue maturity occurs very early (June/July). Managed as winter crops, cereal crops also cover the soil in winter and leave it bare in summer. This situation is very unfavorable for surface microfauna in summer, which cannot tolerate direct sunlight. Moreover, the absence of vegetation during these hot months prevents the use of mineralization products, which inevitably occur, and thus requires the implementation of a winter cover crop to recover these excess nitrates. Nitrogen residues, a source of losses and pollution, often result from the crop/natural climax mismatch. **The corrective option would be to establish a cover with a very short cycle so that it can reach sufficient maturity before frost and return cellulose and lignin. Millet and foxtail millet can be interesting options in this regard because they germinate with very little moisture and have a rapid vegetative cycle (60 days).**

Note: the most important aspect is not the quantity of biomass produced per hectare, but the maturity of the plant at destruction and the synchronization of its growth with the natural vegetation of the climatic zone.

Nitrate pollution: Is wheat to blame?

The continuation of mineralization after harvest, the incomplete utilization of synthetic fertilization applied in spring, the cessation of vegetation out of sync with the climate, and the breakdown of storage mechanisms are all ecological mismatches related to the very cycle of small grain cereals. These mismatches favor poor nitrogen utilization efficiency. A crop like corn (see next paragraph) is much more in harmony with the climate and more effective in nitrogen utilization. It is this climactic mismatch between straw cereal crops and the environment that makes cover crops and "catch" crops (CIPAN: Culture Intermédiaire Piège À Nitrates [nitrate-fixing intermediate catch crop]) a necessity and a significant environmental advancement.

10.1.3 Best practices for incorporating a green manure behind a cereal crop

■ A ration for the soil

The return to the soil of an immature intermediate cover crop is a situation that has become very common and even promoted by modern agricultural practice. During the harvest of the main cereal crop, it is possible to anticipate the destruction of the cover crop, which is still completely immature, by leaving the tallest stubble possible (ideally mowed with a so-called "stripper" header) and seeding the green manure inside it. Once the green manure is destroyed, the simultaneous return of a too-young plant and the standing straw will provide a less unbalanced food source for soil microbiology. The synchronicity of the return of the two materials is important.

■ Green manure that is too young: grazing, a very favorable solution.

When possible, it's preferable to graze the cover crop that is still too young to feed a favorable decomposition flora. As we discussed earlier, passage through the ruminant's digestive system leaves behind materials that are very conducive to the expression of humifying microflora. Rapid-action components, such as simple sugars and non-lignified celluloses, which are so beneficial to bacteria with poor carbon utilization capabilities, will have been digested and utilized by the animal.

10.2 Corn, sugar beet: ecological plants!

Forests establish themselves in climatic zones with rainfall exceeding 650-700 mm/year. An ecological, scientific, and objective perspective attests that many often criticized crops (such as corn) have cycles that closely mimic the primary forests of our regions. Indeed, deciduous trees have active vegetation from spring to autumn, with high water consumption. In autumn, they leave a heavy cover of plant debris on the ground, which has all winter to humify, but also to protect the soil from the destructive effects of rainfall and the "splash" effect of water droplets. Corn, harvested for its grain with the restitution of stalks, offers the same annual patterns of water consumption, restitution of plant residues, and soil protection through its canopy against heavy downpours in the summer and against winter erosion by the plant debris covering the ground. Furthermore, it is a crop that does not require fungicide protection, at least in Europe.

10.2.1 Biodiversity under corn cover can be surprising

To illustrate this assertion with real examples, we will discuss a study conducted over three years with the assistance of an association, Symbiose 03, whose purpose is to associate field measurements with cultivated systems. The protocol was carried out by the National Museum of Natural History in Paris, and the data were analyzed and processed by INRAE in Clermont-Ferrand on about thirty farms operating in identical agrosystems. Although there is homogeneity in situations and measured results, we will focus particularly on a conventional farm on the edge of the Allier River (Maurice L. in Toulon-sur-Allier). It turns out that biodiversity and soil biological activity are higher in the monoculture fields of irrigated grain corn, which has been irrigated for fifty-three years (with the restitution of stalks on the soil), than in neighboring fields of non-irrigated, extensive natural grassland. The biodiversity there is as wide as in natural forest environments in the region.

10.2.2 No nitrate leaching

This farm has seven wells located within its contiguous 300 hectares. Monthly water quality monitoring conducted for over thirty years shows that corn crops along the edge of the Allier River efficiently trap nitrogen. The nitrate levels recorded in the groundwater and in the three wells were 14 mg/liter in 2019 (within a range of 5 to 28 mg/liter, based on all monthly samples taken throughout the year); they are significantly lower than the values found under adjacent extensive natural grasslands used for grazing by Charolais suckler cows. **Corn, like sorghum or sunflower, utilizes all the nitrates released through mineralization since their growth and nutrient requirements are synchronized with mineral availability.** Sorghum, sunflowers, or millet are preferable in areas with limited water resources. The restitution of stubble and stalks, not buried or only very superficially, protects the soil from the effects of winter rainfall and prevents compaction. Their humification on the surface of the soil strengthens the system. Under these conditions, a winter cover crop becomes completely unnecessary and even counterproductive when water resources are limited. A cover crop in winter and spring will consume water that will be sorely lacking during the main crop's growing period. In such situations, we assert that well-managed corn monocultures are environmentally friendly crops that can achieve yields of over 100 decitons/ha. This farm was conventionally tilled for forty-five years and only adopted simplified cultivation techniques in the last five years. On this typical farm in the Allier Valley, with very heterogeneous soils, the white sandbanks have created superficial brown soil after fifty years of grain corn cultivation. Organic matter levels in the sands have increased from 1.4 to 1.8% humus, and in the "stronger" areas, from 3.2 to 4%. All indicators have become positive again: biodiversity, productivity, nitrogen efficiency, absence of pollution, and humus levels. All achieved without cover crops but by respecting climactic synchronicity.

Irrigation: an environmentally responsible approach

> As we've seen, natural ecosystems were primarily based on forests in our regions. Trees had the ability to reach for water deep underground with their powerful root systems. An environmentally responsible approach is to restore this lost function from deforestation. Drawing water from an aquifer that trees would naturally access themselves involves providing the means to ensure summer vegetation cover. This ensures cover and coolness for biodiversity that would otherwise disappear. We enable a diverse fauna of ground beetles, annelids, and springtails to thrive. It's environmentally responsible to create a water reserve to allow an entire chain of life to survive the disappearance of trees.

10.3 Ecological balance of perennial crops

10.3.1 Agroforestry: a system that maintains its own fertility

Agronomic approaches like agroforestry are emerging to restore the functionality of trees in current agroecosystems. These approaches are certainly promising. However, they entail profound changes in the organization and economics of farming systems. It's an ecologically relevant avenue: the vital functions of an ecosystem are partially restored. Moreover, it's one of the situations where the recovery of RCW (ramial chipped wood) meets one of the postulates we had set forth: that all hectares produce their own fertility restoration tools. Here, RCW would be produced and used on-site.

10.3.2 Orchards: the phytosanitary lock to be picked

By definition, agroforestry systems mimic the forest cycle most effectively: they have the same growth, maturity, vegetation cessation, and organic matter restitution cycle in autumn. The major downside in this favorable scenario is the heavy disease protection required in these productions, especially with fungicides. This is where the problem lies because, obviously, the decomposing fungal flora will be affected, and humus-building fungi will be strongly inhibited, whether with synthetic products or with copper-based sprays or certain essential oils. The solution will come when we have effective protection products, such as elicitors, free from these unintended negative effects. We also know that a living soil, in any case, significantly reduces disease pressure. A cover crop will be welcome if allowed to grow at the end of summer to produce lignified tissues. These will provide a choice food for humifying fungal flora as long as these cover crops have been partially spared from fungicides. Indeed, the canopy is sprayed by fogging. Grazing these cover crops, when possible, is the most virtuous solution.

10.3.3 Vineyards: leaving an active summer cover

Its cycle, like that of fruit trees, closely mimics the arboreal vegetation of temperate zones. The difficulty of disease protection is similar to that for orchards. Managing the cover crop through intercropping is particularly interesting to carry out with the vine's own cycle. In order not to compete with the vine in spring and until flowering (the vine absorbs 80% of its needs before this date), it is necessary to control the cover crop through mowing or grazing. Mowing in spring reduces rooting depth and water consumption. Then, let the cover crop grow as it pleases. It will have time to grow, then start seeding, and finally dry out on the vine. This has multiple advantages: its active growth in July/August will allow it to absorb mineralized nitrogen following any heavy downpours in the summer. This will prevent the vine from producing verjuice and save a lot of pruning, while ensuring that the vines are well-matured. The final advantage is to leave mature grasses at the end of the season that can remain on the ground all winter (like dead leaves in the undergrowth), providing fungal flora with a choice food source after a season of fungicides that were not very favorable to them.

10.3.4 Perennial cereals: a source of hope

We are increasingly hearing about research efforts aimed at developing perennial cereal crops. The development of perennial varieties of cereals, legumes, and oilseeds could restore some functionalities of the forest ecosystem by establishing a permanent and even deep perennial cover. American biologist Wes Jackson has been a pioneer in this field, initiating research to this end (he founded the Land Institute in Salina, Kansas, USA). Several varieties have already been developed, including a new cereal called Kernza®. The yields of these "perpetual" crops are still too low to gain significant economic traction. Vegetative period, water requirements... Establishing these crops in temperate regions will still be somewhat misaligned with the local ecology, necessitating careful agronomic management to handle this mismatch. However, they are well-suited to areas where the great prairies once existed in the United States (zones with 400/500 mm of rainfall per year).

10.4 Enabling grassland to maintain its root system

It is very common to observe in our temperate climates that the meadows are sunburnt in the summer. This situation is so frequent that we find it normal. However, when our meadows are sunburnt in the summer, a quick glance around usually reveals that this is not the case everywhere. The grass on the roadsides is often a vibrant green. Cornfields are also lush, as is the forest. Why is the grass in the meadows consistently dry in the summer? In reality, it is entirely possible to align the growth of the grass (within certain limits) with that of the forest. This management involves paying attention to the qualities of rooting.

10.4.1 The impact of grass cutting on root systems

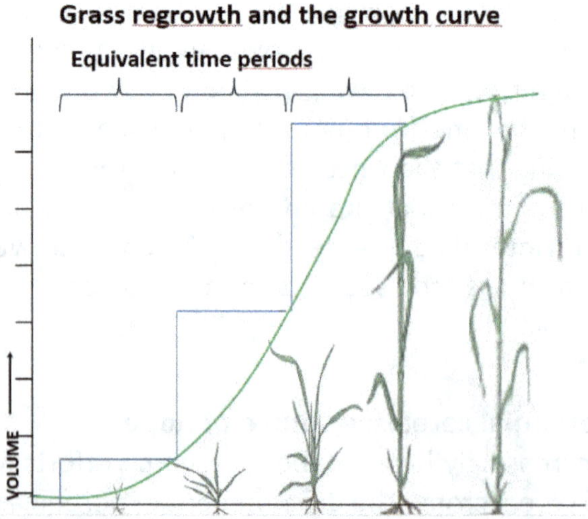

Figure 10.1 – Grass regrowth and growth curve

The first point of attention in grass management is to carefully consider that cutting the grass removes part of its photosynthetic capacity. If the cut is too short (less than ten centimeters in height), the time it takes for the plant to regenerate its "solar panels" will be long. Consequently, growth will be hindered for two to three weeks before resuming at a rate that can then reach 250/300 kg of dry matter per day. However, there is something even more detrimental to the meadow.

When the aerial part of the plant is excessively amputated, the plant can no longer sustain a complete root network. This will cause significant root mortality and reduce the depth of soil exploration. There will then be a cessation of root exudation in the deep subsoil, a halt in the feeding of the associated microflora, and ironically, even though the plant has reduced water needs, it will immediately suffer from drought because it no longer has the adequate rooting to draw deep water. This recurring phenomenon causes the "absurd" situations mentioned earlier, where we see green roadsides in summer because they are not mowed, and burnt meadows because they are cut too short. Therefore, to maintain summer growth, it is imperative not to overgraze and to practice very rapid rotation of the animals: one day, one plot, a daily change. This is the best way to avoid imposing regular mowing on certain areas of the meadow, which would exhaust the plants. Thus, it is the best way to ensure grass production throughout the season.

“Imitating” the instinct of large herds

Large wild herds of hundreds of thousands of animals have an instinct for proper pasture management. They never stay in one place for more than a day. There is no waste because what is not consumed is trampled due to the high density of animals (mob-grazing/Allan Savory), encouraged by the fear of large predators (lions, wolves, pumas, etc.). The grasses maintain a sufficient height to sustain their complete root system. The herds never return to the same spot before six to eight weeks, which allows them to escape parasitism (pulmonary strongyles, worms, helminths) whose cycles average 21 days.

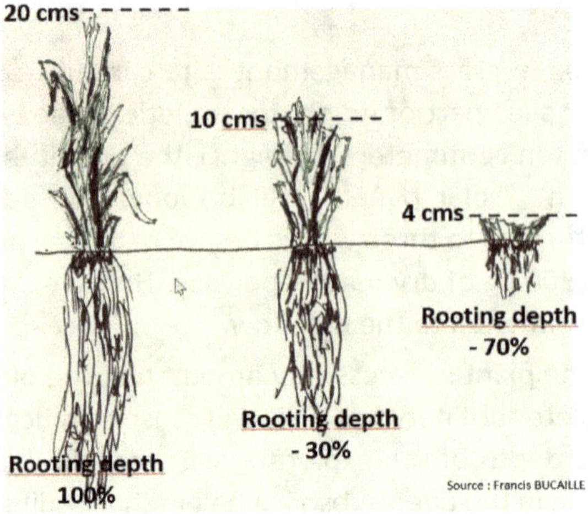

Figure 10.2 – Impact of cutting height on rooting depth and productivity.

10.4.2 Planning for pearl millet or grazing corn in the event of drought

Fine grazing management, as we recommend, is not a 100% guarantee of achieving summer growth. In years where drought is prolonged, the meadow will still struggle to maintain its productivity. The critical mistake to avoid in such situations is overgrazing. It is essential to know when to "disengage" the meadow, leaving it at a height of 20 centimeters. Even if it looks hay-green or brown, it will have the capacity to resume full production once the rains return. In the meantime, grazing on pearl millet or forage maize (special varieties that tiller) should be done, which would have been planned for this purpose. As mentioned earlier, these summer plants are well-suited to the climate.

Overgrazing causes nitrate leaching

In summer, when the pasture is "shaved," its depth and root density become severely atrophied due to lack of photosynthesis. However, the soil, exposed to the direct heat of the sun, increases in temperature and maintains a slowed mineralization in the cool semi-deep horizons. As soon as the first rains occur, the soil is ready to resume mineralization in all horizons. The phenomenon explodes even before the pasture's root systems are rebuilt. This root reconstruction takes a lot of time as it starts from scratch. Leaching becomes inevitable in this situation.

10.5 Active summer vegetation: a collective ecological benefit

■ Bacteria that "make the rain fall"

It had already been observed that bacteria could be transported over long distances. Some had even demonstrated that millions of bacteria were deposited on the ground after severe storms. However, German researchers analyzed 42 large hailstones from a storm that hit Ljubljana, Slovenia, in 2009. It turned out that the number and density of bacteria present in the hailstones were equivalent to what is found in river waters. Many of these bacteria are equipped with brown pigments that allow them to resist high-altitude UV radiation. Scientists believe it is not coincidental that such significant quantities of bacteria are found in storm rains and hailstones. They are beginning to gather well-established evidence that clouds cause violent updrafts, and **bacteria present on leaves are carried into the heart of clouds, initiating rainfall.**

□ No plants in summer, no epiphytic bacteria, no rain...

Clouds are not enough to make it rain. Water vapor droplets need to agglomerate. For droplets to agglomerate, they need a nucleus to trigger nucleation. The droplets then grow in size, become heavier and reach a critical weight. Then, despite the ever-present updrafts in these clouds, they fall: it rains! Bacteria present on plant leaves, rich in pigments to protect themselves from UV radiation, are abundant on vegetation leaves and have exactly the same characteristics as those found in hydrometeors. The obvious condition is that there must be vegetation from July to September. In addition to the evapotranspiration generated and thus the cloud formation, active vegetation transforms these clouds into rain. This interaction, long suspected, turns out to be a common and significant phenomenon.

How inspiring it is to think that the Gaia theory (James Lovelock) may find a demonstration of its correctness here. Bacteria are drawn into the sky by a benevolent wind to provoke rain, which will nourish a canopy that will host the next generations of bacteria. Respecting climactic synchronization (of vegetation in midsummer) thus has a positive influence not only on the soil but also on... the sky.

□ Millions of hectares biologically dormant in summer

Millions of hectares of stubble where the establishment of vegetation cover is uncertain and climate-dependent. Millions of hectares of sun-scorched meadows that expose bare soil to the sun's rays. All of this can only cause suffering to surface microfauna. All of this can only raise the ambient temperature. All of these biologically dormant surfaces during the summer do not contribute moisture to the atmosphere through evapotranspiration and deprive clouds of living nucleation nuclei. What a pity to let all this water evaporate in winter through vegetation covers when it could produce so many beneficial effects in summer. The ancients had the intuition: "if you want water, take care of the forest." It is no coincidence that the land managers of centuries past created the department of "Waters and Forests" (governmental organization to manage the rivers and forests in France). The two are intimately linked.

Reducing climactic lag: the new reservoir of productivity

> Even though climax ecosystems have become rare today, it is extremely interesting for farmers to refer to them. The inherent capacities for exchange, recycling, storage, absorption, evapotranspiration, disease control, and productivity are at their maximum, all without any external intervention. Observing the climax thus gives everyone the opportunity to imagine the extraordinary potential that lies ahead, knowing that much of the functions diminished by soil domestication can be reactivated. The profession of the farmer will increasingly have to integrate this mission of reducing or limiting the negative effects of ecological mismatch with the climax ecosystem. This is where the reservoir of productivity lies, compatible with the economy and profitability of the farm, compatible with societal expectations, compatible with the planet, compatible with human health, and compatible with life.

10.6 Decompacting minds to change paradigms

Man must abandon the vision of himself placed at the top of the pyramid of life. In this chapter, we will reiterate this "ultimatum" that we presented in the preface of this book. Man has everything to gain from moving away from this vision. We believe that this is the only way for him to truly understand the importance of putting the original climax ecosystem into perspective. Otherwise, the divorce from the soil will continue to occur.

10.6.1 Our mental projections drive our practices

This development may seem off-topic because everyone thinks that agronomy and agriculture are merely cultivation techniques, plant breeding, and fertilization recipes to be applied to the soil and plants. In fact, we never arouse more interest when we give lectures than when we give a recipe for foliar application or for maximizing the effect of glyphosate, in order to reduce the dose and therefore the impact on the soil. You are right, agronomy is 99% technique and 1% cultural. But we are

not off-topic at all because it is this 1% of cultural and collective unconscious (cf. C.G. Jung's archetypes) that drives the other 99%. This 1% channels the orientations of research, development, and production.
For example, let's take the history of INRAE, whose genealogy is quite typical of the last two ages of agriculture, as we have described them: in 1955, the creation of the public establishment CNEEMA (Centre national d'études et d'expérimentations du machinisme agricole [National Center for Agricultural Machinery, Rural Engineering, Water, and Forests]), which became the CEMAGREF in 1981 (Centre national du machinisme agricole, du génie rural, des eaux et forêts [National Institute for Research in Science and Technologies for the Environment and Agriculture]), then IRSTEA (Institut national de recherche en sciences et technologies pour l'environnement et l'Agriculture [National Institute for Research in Science and Technologies for the Environment and Agriculture]). Finally, by merging with INRA, it becomes, in this remodeled research, INRAE (Institut national de recherche pour l'agriculture, l'alimentation et l'environnement [National Institute for Agricultural Research and the Environment]). We can clearly see this shift from pure machinery to food and then the environment with all these intermediate stages. The semantic shift and the objects of their missions have indeed evolved. Do you think that these organizations precede the collective unconscious, of which public opinion is just one facet? No, they are a direct product of it. The cultural and mental representations take precedence.

10.6.2 Culture governs agriculture

Today, on the same planet, we see France mocking Kervran[12] with his biological transmutations, immolating a Benveniste with the memory of water, and India valorizing its homeopaths with 100 CH dilutions applied to agriculture. You have American farmers who, without hesitation, laser-level their fields for flood irrigation, and Waipi tribes in South America who, after apologizing to the spirits of the forest, practice very controlled slash-and-burn agriculture, which likely gave birth to the famous "Tera preta" soils that are so fertile. Simultaneously, in the same Western part of Europe, in 2020, you have Portugal planting vineyards by stripping the topsoil to replace it, after excavating hills with heavy machinery for several meters and filling in excavations; and at the same time, France prohibits any transport of soil or any soil disturbance operations whatsoever to protect its sanctified vineyard terroirs.

---

[12] Corentin Louis Kervran (1901 – 1983) was a French scientist. Kervran proposed that nuclear transmutation occurs in living organisms, which he called "biological transmutation"

10.6.3 Finding the right rudder

Through these few examples, which could be multiplied indefinitely, we understand that **it is humanity's projection in its relationships with the Earth and with the soil, it is the cultural and educational legacies that will dictate tomorrow how to cultivate**, plow, with religious devotion or not, to respect the natural laws of the climax or not, to ensure that crops can do without fungicides or not, to replenish humus stocks or not. There is no doubt that the technical solutions are almost all already there. It only requires individual wills and collective decisions to pave their way between trends and habits. The third age of agriculture will be golden if we decide to make it so, or at least it will be tomorrow what we decide to make it today.

D

A global approach to soil: the soil profile

We have seen that the major soil functions are at their peak in the presence of climactic equilibrium. To give plants cultivated more competitiveness, the "ecological" farmer will thus have to try to restore a set of these natural functionalities that have been weakened, degraded, or put to sleep. To achieve this, one must first know his or her soil, its functioning, its strengths, weaknesses, and potential. What the French poet and writer Victor Hugo expressed in a lyrical form "I was reading. What was I reading? Oh! the old austere book, the eternal poem! — The Bible? — No, the Soil!" (*Les Contemplations*, V. Hugo), we will try to do here with a synthetic, comprehensive, pragmatic, 3D soil profile approach updated in light of current knowledge.

This approach combines short-term investigation (finding traces and consequences of agricultural practices and remedying them) and long-term considerations (pedogenesis and its consideration) based on indicators from four complementary sciences (biology, physics, chemistry, hydraulics). It also proposes to estimate reversible and irreversible effects caused by natural soil aging as well as those generated by human action.

11 • SOIL FORMATION AND SOIL PROFILING

11.1 Soil formation: by living beings, for living beings

11.1.1 Soil history: from pedogenesis to pedology

All soils are the result of the evolution of a mineral substrate. Most soils are neoformed from the parent rock, meaning their mineral fraction originates from it. The parent rock can be of sedimentary (chalk, limestone, sands, windblown silts), igneous (basalt, granite, lavas, etc.), or metamorphic (schist, gneiss, etc.) origin. However, soils formed from aeolian, terrestrial or marine alluvial deposits such as polders, or glacial moraines are exceptions that deserve consideration in the agronomic approach to these soils and lead to different management of surface fertilizer and amendment inputs.

What transformed these inert materials into a complex environment favorable to multiple forms of life is life itself through evolutionary processes and thanks to simple unicellular organisms, sometimes extremophiles at first, to an ecosystem with multiple layers, as we know it today. We will first examine some of the key mechanisms that have contributed to the genesis of soil.

11.1.2 The crucial role of bacteria

Soil is a bit of mineral and a lot of living matter! All forms of humus (mor, moder, mull) are the result of soil life activity. But beyond organic matter, it appears that even the most abundant, yet complementary to humus, mineral matter, namely clays, is the result of microbiological activity. Today, we know that the production of clays from parent rock materials is accelerated thanks to bacterial activity (Gui Li, 2019). Research teams have demonstrated that within a few days, bacteria are capable of creating kaolinite-type clays from inert volcanic or sedimentary materials (Saverio Fiore, 2011).

11.1.3 Earthworms

Earthworms of the anecic genus have a very important soil regeneration capacity. They are active in temperate climates for about 200 days per year during which they produce their own weight in casts every day. One ton of earthworms per hectare can vertically move around 200 tons of soil per hectare per year. Each time, they bring back from the depths the clays lost by leaching and cations (calcium, magnesium, potassium, etc.). Thanks to their so-called "Morren glands," which act as "mineral solubilizing factories," including calcium, anecic earthworms can bind clays and humus together. They effectively rebuild CEC (cation exchange capacity)! However, in temperate climates, as well as in tropical and semi-tropical regions, some soils, especially siliceous sandy soils, are not suitable for hosting anecic populations. In these cases, termites or ants often efficiently ensure soil regeneration through bioturbation. A first conclusion is that earthworms cannot be the sole criterion for assessing soil fertility.

Darwin and earthworms

It's a little-known fact that Charles Darwin harbored a "guilty" pleasure in the scientific observation of earthworms for half of his life. So much so that when he prepared to publish his book "The Formation of Vegetable Mould through the Action of Worms" (vegetable mould being more commonly known as topsoil) in 1881, he took care to inform his mentor, Professor Carus, in these words: "I have now sent to the printer the manuscript of a little book on the formation of vegetable mould by earthworms. It is a subject of little importance, and I do not know if it will interest any readers. But it has interested me." In fact, what started as a small book of little importance became a major work. Bioturbation, fertility transfers, burial of organic matter in depth, mineral recycling, grinding of rock fragments, shaping of landscapes, etc. In his work, he described with much foresight almost all the functionalities of earthworms that we know today.

11.1.4 Clays

Alongside the bioturbation carried out by earthworms, ants, and termites, another mechanism capable of restoring soil structure is linked to the structural activity of clays. With their layered structure, so-called "swelling" clays (illites, montmorillonites) have the ability to swell with water. During dry weather, as water is withdrawn, the clays contract, and cracks appear. This phenomenon is purely physical and depends solely on the nature and quantity of clays.

11.1.5 Some soils are particularly fragile

These are soils composed mainly of siliceous sands, sandy loams, or sandy loam soils (area overlapping the zones delimited by the solid line and the dotted line). This type of soil can be identified using the texture triangle below, based on the clay content (ordinate) and values of sand (abscissa). By adding clays and sands and subtracting from 100, we obtain the percentage of silt and loam (100 – (40% clay + 15% sand) = 45% silt/loam).

These soils are not suitable for the development of anecic earthworms, as they are too sandy and lack the other restructuring factor, which is clay. Note that the sands affected by this latter consideration are non-polished siliceous sands (anecic earthworms do not have the same aversion to limestone sands). These soils, which are particularly at risk, require special attention. No-till or minimum tillage techniques are much more delicate to implement here than elsewhere and may require occasional use of a subsoiler based on diagnosis.

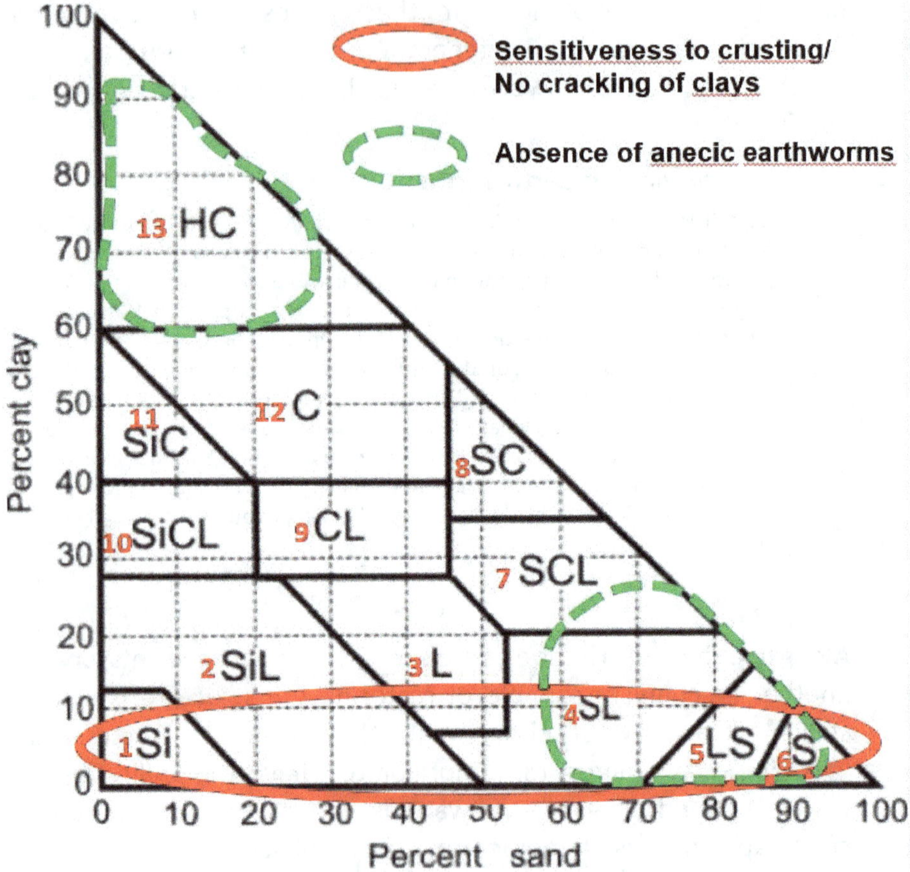

Figure 11.1 (left)– Texture triangle. Left : Ranges of susceptibility to crusting and absence of anecic worms. Right : Readily Available Water according to soil texture.

In their original climax (tree vegetation), these soils benefit from the contribution of large trees: a dense, permanent, and voluminous rooting system of the large trees, which prevents the soil from collapsing on itself. In these situations, agronomic plowing always makes sense and can partially compensate for the functions destroyed by cultivation: periodic physical restructuring and the upward movement of cations.

Agronomic Plowing

Agronomic plowing is the plowing practiced by our ancestors. It is a plowing that should not exceed a depth of 12 to 15 centimeters with furrows no wider than 12 inches, in order to achieve proper soil turnover without the need to work deeply. No skimmers are used, to avoid pressing fresh organic matter into the bottom, and a large number of furrows are used to allow for off-furrow work.

Benefits of shallow plowing: fuel savings, mineral recycling (including calcium), combating surface acidity, preserved effectiveness of root herbicides, and prevention of plow pan formation.

11.2 Exploring the third dimension of soils with a soil profile

11.2.1 The benefits of a direct 3D approach

Soil analyses, conductivity maps, yield maps, presence of castings... We are often content to study agricultural soils indirectly and superficially (soil analyses, penetrometer etc.) or in two dimensions by observing the soil surface and mapping. There are two pitfalls here. Firstly, anything based on indirect measurements does not correspond exactly to reality, but is already an interpretation. Secondly, we don't just cultivate surfaces expressed in hectares, but also volumes. We don't adopt the same practices when we plan to cultivate a hectare on a soil column of 30, 50 or 150 cm! The first 30 cm represents around 4,500 tons of soil per hectare. However, annual crops have an exploration potential that can climb to 10,000 tons, 20,000 t, 25,000 t/ha, or even much more for perennial crops such as alfalfa! Gains in autonomy, efficiency and ecology largely depend on this new approach, which takes into account all levels of the soil.

11.2.2 Digging a soil profile

■ Methodology points

- **Depth.** The primary purpose of a soil profile is to observe the functioning of the soil down to the root limit. The soil profile should be dug at least to this limit. In temperate regions, this can reach, for annual plants, two meters or even a bit more.
- **Length.** The length of the profile should be at least 3 to 4 meters and it should be made perpendicular to the direction of tillage. This is to reveal the impact of the circulation and work of agricultural equipment and to conduct an observation that is as representative as possible of the plot. A sufficient length is also necessary to limit the risk of observing a section

that is too particular (a soil that has already been deeply disturbed in the past, localized compaction, etc.).

- **Orientation.** The presence of cast shadows on the profile wall illuminated by the sun hinders observations. Ideally, the face chosen for study should be the one that will remain in the shade at the time of observation.
- **Scraping.** The profile should be scraped with a knife. To achieve this, one should proceed from top to bottom diagonally with a quick rotation of the blade. This way, the debris removed falls onto the parts of the profile that have not yet been refreshed.
- **Farmer involvement.** The most important rule is certainly that the farmer feels involved in the observation of the soil profile. Creating a soil profile is a scientific agronomic experience, but it is also a sensory experience. When one puts their nose in their soil, they are often surprised by what they find: colors, fauna, root depth. It is the only way to observe the consequences of past mechanical interventions and to imagine the most suitable future technical routes: soil tillage, fertilization, irrigation, and drainage.

Accessing soil memory

> From an agronomic perspective, the main interest of this exercise in consulting the soil's memory is to find traces of the pre-existing climax ecosystem and to identify which functionalities can be restored and to what extent. It is the only way to identify the margins for improvement.

■ Necessary equipment

□ Essential

Backhoe, meter, knife, hydrochloric acid, pH meter (Hellige type), large bottle or canister of water, notebook, 20x binocular magnifying glass, blower...

□ Optional

Potassium thiocyanate, infiltrometer or Guelph permeameter, Munsell color chart...

11.2.3 Defining soil type

■ Horizon delineation

The first observation that stands out when opening a soil profile is the chromatic differentiation of the various horizons, schematically represented below. In general, very gradual contrasts can be interpreted as a sign of effective bioturbation and mixing of the different horizons. The more gradual the continuum, the better the potential of the soil will be utilized.

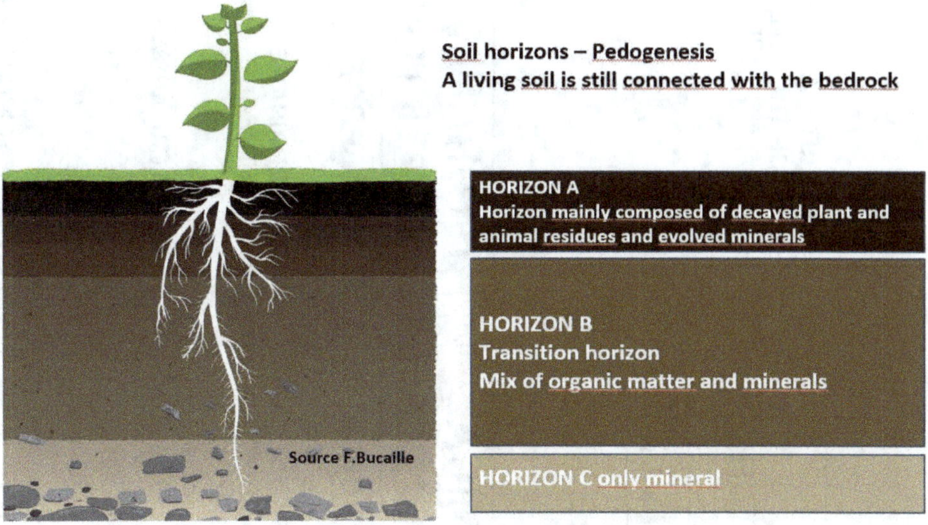

Figure 11.2 – Soil horizons - Pedogenesis. A living soil is still connected to its parent rock (source: F.Bucaille)

■ Chronology and nature of descriptive elements

Agricultural soil derives most of its fertility from the clay-humus complex, an association of two materials that are theoretically opposed. Indeed, they both carry negative electrical charges and should therefore repel each other. However, the bridges provided by divalent minerals such as calcium, magnesium, or iron ensure stability to the whole. The organic matter (OM) of the soil originates from living tissues, which is quite understandable. As we have seen previously, the genesis of clays also involves a biological process. Therefore, we will first examine biological activity and then other compartments of fertility:

- Biological fertility: rooting depth, earthworms and their depth, presence of macrofauna, evolution of stubble and harvest residues.
- Physical fertility: arrangement of soil aggregates, surface structure, and plow pans.
- Chemical fertility: carbonation, surface and subsurface acidity, redox phenomena, and gley.
- Water fertility: water storage, infiltration capacity (hydromorphism?), and water retention.

This segmentation of fertility components is, of course, artificial because, as we will see, soil microbiology, physics, chemistry, and hydraulic functioning have complex interactions.

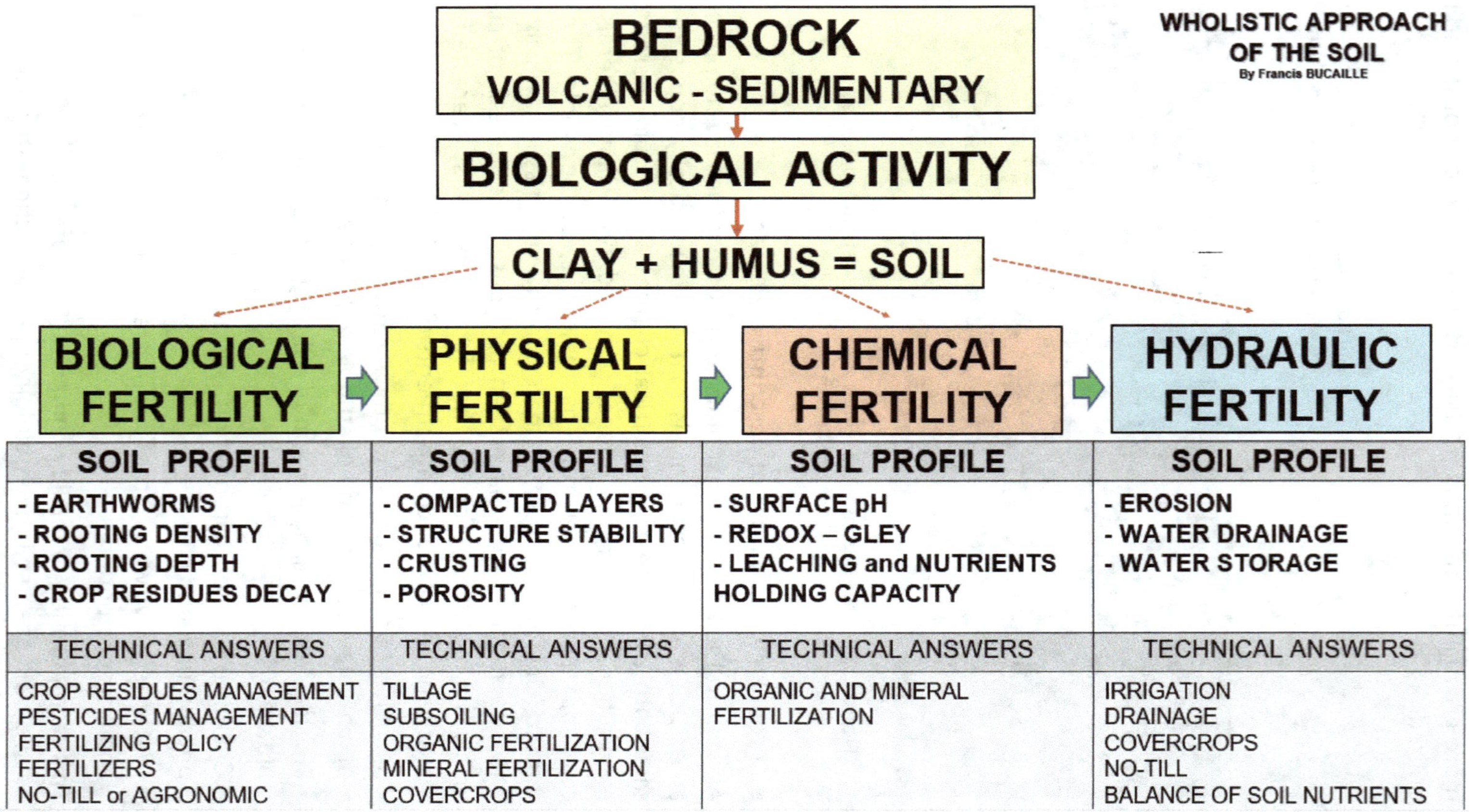

Figure 11.3 – Global approach to soil

12 • BIOLOGICAL FERTILITY ANALYSIS

Earthworms, roots, bioturbation, mycelium, etc. Comparing ancient traces of life to present ones allows for the qualification of discrepancies. These differences are rich in insights into soil dynamics and the maximum potential of processes, differences from which the farmer can set achievable goals. In the field, while meeting with farmers, we encounter discourses that are both overly optimistic about new practices and overly pessimistic about the current state of affairs. Thus, in this section, we will learn to:

- assess the current situation;
- set realistic improvement objectives.

12.1 Assessing macrofauna activity: ants, anecic earthworms

12.1.1 Challenge: anecic earthworms play a major role

Anecic earthworms are large-sized worms that live in vertical burrows. When present, they traverse the entire soil profile. The presence of anecic earthworms is a major factor in the biological fertility of soils. The first thing to check during soil profiling is the current or past presence of these anecic earthworms. In some soils, ants or termites fulfill this role. Not to mention the natural work of larger fauna, such as moles, in the environment.

12.1.2 Observation method

Identifying active burrows. The observation of anecic earthworm populations is indirectly conducted during soil profiling by observing their burrows. A burrow is considered occupied and active when the burrow wall is shiny and covered with a thin mucus that facilitates the worm's movement, as well as root growth. This film lining the burrow walls contains auxin-like growth substances and nitrogenous materials that can double the growth rate of roots and root hairs. This point is crucial because it allows, during the establishment of a main or cover crop (whose objective is to act as a catch crop for nitrates), the plant to win the speed race initiated by the first rains: to catch up with nitrates as quickly as possible to recover as many as possible. Live roots and active earthworms frequently coexist. Deserted burrows, on the other hand, show signs of decline. The wall is no longer covered with mucus. One also observes traces of filling by colored materials consisting of millimeter-sized fecal pellets and excrement from secondary fauna that settles, as well as traces of leaching of clays, silts, and humus from the upper horizons. Leaching traces are always darker than the soil itself. Sometimes old burrows are completely filled in. The soil filling them has very little cohesion with the rest. This is a sign confirming that it is indeed an old burrow, which can easily be detected using a knife. Complete filling indicates a period of inactivity of the earthworms spanning several decades. Since the burrow initially retains its functionality for water drainage, it may take some time before the impact of agricultural practices leading to the disappearance of these "soil engineers" is noticed. It is noteworthy that a constantly maintained burrow, lined with the shiny film produced by the earthworm, not only allows water to infiltrate rapidly into the deep horizons but also does so without carrying away any soil, silt, or clay particles, thus avoiding any leaching of fine elements.

12.1.3 Population assessment method

- **A 20 x 100 cm plot laid horizontally in the heart of the compacted plow pan.** The assessment of earthworms is done indirectly by counting occupied and unoccupied burrows. Depending on the time of observation, the number of active earthworms fluctuates greatly, so much so that it seems more effective to count the number of burrows. **We recommend conducting this count horizontally over an area of 1/5th of a square meter, distinguishing between burrows with a diameter smaller and larger than 3 mm. This distinction will allow for differentiation of the present species, larger anecic and smaller endogeic worms. It is desirable to carry out this count in the most compact horizon because it is the most limiting horizon**, but also the one that allows for the observation of the effectiveness of the work done. There are situations, as frequent in plowing as in direct sowing, where significant earthworm activity can be observed on the surface without it being expressed deeper. A living soil is primarily characterized by the fact that surface horizons are always connected to deeper horizons. The highly useful function of earthworms is to recycle and mix the horizons: bringing clays to the surface while combining them in their intestines with fresh organic matter. The product, in the form of castings ejected onto the surface, is practically pure CEC; Morren's glands even construct the calcium bridge. These worm casts exhibit astonishing structural stability and can resist months of rainfall. This bioturbation work also helps to transport plant debris deeper into the soil.

- **Counting and calculating the occupancy rate.** The sum of active and inactive burrows gives us the maximum number of burrows achievable by the soil under optimal functioning conditions. Ideally, soil functioning at its optimum has an occupancy rate of 80-95%. Observing the occupancy rates of the burrows also informs us about the dynamics of earthworm populations. The reference should be the number of burrows that existed at its peak in the semi-deep horizons (beyond 40 centimeters in depth, below any potential plow pan).

  (100 × number of occupied burrows)/Total number of burrows

  = **Occupancy rate**

12.1.4 Analysis: dynamics to be highlighted

- The soil profile allows us to reveal processes of disappearance, absence, decline, improvement, and optimum conditions.
  - **Disappearance.** A soil that shows no signs of active earthworms but contains unoccupied or clogged burrows indicates the more or less recent disappearance of these species. A complete analysis of the soil profile and the historical practices on the plot often allows identifying the causes before establishing a restoration plan: less aggressive soil work, shallower plowing, cessation of the use of aggressive substances like certain slug repellents, essential oils, copper-based mixtures, etc.
  - **Natural absence.** In some cases, the soil shows no current or past traces of burrows, even in the deepest horizons where the soil memory dates back several millennia. These soils are devoid of any anecic activity and cannot naturally host them. This is the case for siliceous sandy soils, some hydromorphic soils, or plastic clay soils (see texture triangle). However, human intervention, such as drainage in hydromorphic soils, can sometimes enable these soils to host earthworms, even if they never had them before. Apart from these particular cases, it is unrealistic to think that agricultural practices, even sustainable ones, can outperform what nature has achieved over millennia in terms of deep exploration by anecic earthworms. Through massive additions of ramial chipped wood (RCW) or green manure, significant populations of epigeic earthworms like *Eisenia foetida* may appear, although they will remain on the surface.
  - **Optimum, decline, or improvement.** An occupancy rate of 80 to 95% in the compacted horizon indicates that earthworm populations are at their optimum. An occupancy rate below 70%, with freshly abandoned burrows (i.e., without mucus but not yet clogged), is a sign of a population decline in progress.

Are circulation pathways optimized?

Observing earthworm burrows also warrants examining their pathways. Ideally, the various soil horizons should not impede the movement of earthworms (see diagram). If bypasses are observed, it indicates that circulation is not optimal. Some horizons are poorly traversed. Sometimes, they are not traversed at all, confining worm activity to the surface. The soil regeneration process is stalled. A functioning soil is always connected to the parent rock through various processes.

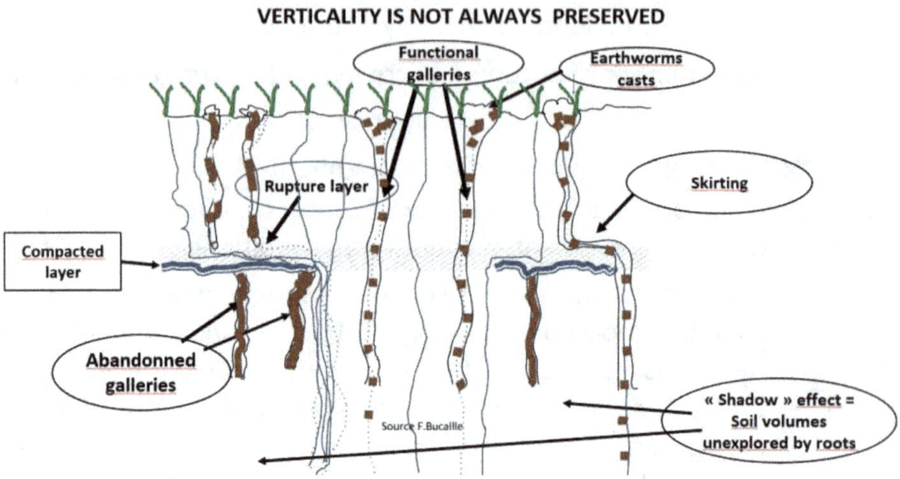

Figure 12.1 – Schematic diagram of anecic earthworm galleries and soil occupation by roots

Counting worms using the "mustard" technique is insufficient

Using mustard as a method to count earthworms is sometimes employed. It involves dispersing a solution containing mustard over the soil, which prompts the worms to surface. While this method may have some educational value, its technical conclusions are limited. Firstly, earthworm activity varies with seasons and weather conditions; for instance, anecic worms have the ability to enter diapause. Secondly, the number of earthworms found on the surface does not indicate their presence at depth or their ability to traverse all soil horizons. Those in deeper layers remain unaffected by mustard. Yet, the most remarkable regenerative function of earthworms stems from their capacity to recycle elements from deep within the soil. Likewise, counting the number of castings on the surface provides a rather unreliable indicator that can even be misleading. Sometimes, anecic worms are trapped on the surface by a plow pan. Consequently, they shuttle back and forth within a restricted volume, unable to "find their bearings." This explains why numerous castings are often observed on tractor wheel tracks.

Collecting castings from a one-square-meter area and weighing them can be a reliable means of assessing earthworm activity. This evaluation is not dependent on their highly seasonal activity.

12.2 Other types of bioturbation

In soils where earthworms cannot operate, two other actors can take over. They should not be overlooked as they are major players and sometimes the only ones on nearly 30% of agricultural land:

- Ants, due to their large colony numbers, are capable of moving very significant volumes of soil, even in temperate climates when soils are not disturbed by mechanical operations. From this perspective, ants are even more sensitive than earthworms to repeated cultivation operations. This is why they are more frequently found in grasslands. Ants always settle in very healthy, non-floodable soils. Sands or sandy loams do not deter them. Like earthworms, they use microbiology to make certain nutrients more assimilable. Some are even capable of cultivating fungi. Termites in warm climates perform even more impressive "civil engineering" and excavation work.
- Dung beetles exist in all latitudes and, as their name suggests, mainly accompany ruminants and are essential in the ecological chain linking the pasture, the ruminant, and the soil. Australia experienced this firsthand after the first settlers imported ruminants without dung beetles. To prevent the ecosystem from being overloaded with poorly decomposing dung, they had to introduce these small animals to close the loop. They can dig up to 30 centimeters deep, digest the feces of other herbivorous animals, and accelerate the process of humification and return to the dung production cycle, while maintaining surface porosity in soils.
- Foxes, rabbits, moles, and mole rats also contribute to soil macroporosity and horizon mixing. Some studies seem to demonstrate that the decrease in grassland productivity caused by the destructive action of mole rats is compensated by yield increases in subsequent years, generated by rejuvenation of grass species and a revival of biological activity

12.3 Analyzing rooting qualities

12.3.1 Challenge: volume and speed

Rooting qualities are crucial for two main reasons. The first, and perhaps the most obvious, is closely related to the volume of soil explored in a functional manner. This volume constitutes the direct fertility reservoir that the plant will have access to (not to mention indirect reservoirs through mycorrhizal exchanges). The second reason is that the quality and kinetics of rooting are also crucial in the crop's ability to win the race against the leaching of fertile soil elements.

12.3.2 Morphological method and analysis

- **Root system architecture** is crucial. Ideally, roots should exhibit a well-branched structure without gaps. They should also not show signs of deviation (crankshaft shape) and must have absorbing root hairs in all soil horizons. Deviating roots indicate a loss of plant competitiveness in its environment, often associated with the presence of compacted zones. The absence of roots can also result from lack of oxygen and hydromorphy.
- **Absorbent root hairs.** Plant nutrition functionality relies on absorbent root hairs. Their presence indicates that this function is well established. Conversely, their absence at different levels of the root system indicates that plant nutrition is not operational there. In such cases, the root function is limited to transportation. Often, the most compact horizon is not explored by absorbent root hairs. Re-establishing absorbent root hairs at all levels of the root system is a medium-term achievable goal in all soil types, with potentially significant benefits.
- **Other signs: number, color, shape, and coiling.** Healthy roots are typically white (occasionally black for certain species like lavender, horsetail, trees, etc.). Yellow, flattened, or fishbone-shaped roots indicate compression and poor functioning of the root system. Root coiling is also common in mechanically cultivated soils, forming air pockets between soil blocks. These coiled roots reveal the presence of soil cavities acting as root traps. Such compartmentalization is detrimental as it represents a significant expense for the plant with limited return on investment.

12.3.3 Rooting kinetics method and analysis

To analyze the kinetics of root development, it's interesting to conduct soil profiles when the cultivated plants are growing. A wheat plant established for two months in a well-functioning soil should have root depths of at least 60 to 75 cm. On average, for annual crops, root growth levels should reach 1 cm per day during the first two months of establishment. It's worth noting that in soils hosting anecic earthworms, vertical galleries act as true "highways," allowing roots to double their growth rate.

12.3.4 Rooting depth method and analysis

The roots of annual plants form plant tissues that always break down very well in the soil. The C/N ratio of roots - between 30 and 40 (80 to 100 for aerial parts) is in fact very favorable to decomposition, whatever the environment in which these tissues are found. What's more, they're bound to be found where there's oxygen, otherwise they wouldn't have grown here. It is therefore more difficult, but not impossible, to identify the rooting history of annual crops a few months after harvest.

Analyzing rooting depths is easier to carry out in any season for perennial plants. We will then focus on observing the maximum depths of live roots and traces of previous root systems (dead woody roots). Dead roots are easily identifiable as they are dry, often brittle, and devoid of sap. The difference between the maximum depth of dead roots and the maximum depth of live roots reveals the exploration potential achievable by the perennial crop in a well-functioning soil, but also exposes all the lost potential (a common situation in vineyards). Similarly, we can also measure a difference in maximum root diameter at each root level between old and live roots.

The rooting depth considered is the one where we still have eight to ten roots per linear meter. An isolated root descending lower remains an exception and cannot be considered indicative of rooting depth.

The estimation of the deep volume historically explored by old traces (root walls loaded with tannins and still visible), voids and more or less plugged galleries are all signs that allow us to set a reasonable target for regaining lost potential. For every decimeter gained, it's 1,300 tons of soil recovered!

Never without oxygen

> Roots never explore areas where oxygen is absent. Their presence is therefore in itself a sign of its presence..

Rooting kinetics and earthworms

> The mucus secreted by anecic earthworms along the walls of their burrows acts like an auxin-type hormone, stimulating the root growth of plants. Inside an occupied gallery, the growth rate of roots is doubled. This phenomenon results from the co-evolution of plants and earthworms. In soils with present and occupied galleries, the establishment rate of crops is significantly enhanced. For instance, the anti-leaching functionality of a CIPAN (catch crop for nitrates) is more likely to be effective, making the agroecosystem more efficient and less dependent on nitrogen inputs.

12.4 The stabilizing role of pseudomycelia (whiskers)

12.4.1 Challenge: pseudomycelia fix cations, prevent surface acidification and buffer the soil

Pseudomycelia are still largely ignored in the field. Yet we are convinced that they play an absolutely major role in fertility in many types of soils where calcium is abundant and presents two simultaneous risks: the loss of bases, and thus surface acidification, and the accumulation of this calcium around the roots, causing nutritional disorders such as iron chlorosis. Indeed, biological activity, and more particularly fungi, produce oxalic acid which binds to calcium to form oxalic acid crystals, insoluble in water, that precipitate immediately around the mycelium. It is the establishment of pseudomycelium by the fungal flora that, in the climax forest system, slows surface acidification despite the decaying process of dead leaves and branch residue (with the exception of forests of certain species such as conifers). These mycelium-origin formations also contribute to fertility in calcareous stony soils by increasing the horizon's capacity to retain water (cf. Gabriel Callot, 1999). By neutralizing excess minerals in a soil (Ca, Fe, Mn...) through weak organic acids (oxalic, lactic, acetic acids...), a very active microbiology buffers the excesses – of pH among others – as they would manifest if the geology were left to itself. This is why soil with weak or unbalanced biological fertility will tend to have extreme pH levels (5 or 8.5).

12.4.2 Method and analysis: detecting traces of pseudomycelia

Traces of pseudomycelium materialize in the form of white veils visible to the naked eye, resembling diffuse cotton or beaded filaments. They are generally located between 10 and 50 cm deep. Observation with a binocular magnifier (×20) reveals that these pseudomycelium filaments are composed of crystals. It is then very easy to determine whether they are organic tissues or mineral formations. The crystals of pseudomycelium are most often grains of calcium oxalate (or even iron or magnesium oxalate in some soils).

12.5 Evolution of crop residues

The sign of intense and balanced biological activity is reflected in the rapid decomposition of crop residues: straw and chaff. There is a general rule, with a few exceptions, that plant products take as long to decompose as they took to produce. Corn cobs are among the exceptions because they contain more lignin and cellulose. The significant presence of discernible plant debris over a year old is a sign of poor oxygenation and defective microbiology.

The biological examination of the soil profile also involves an olfactory examination. Bad odors are signs of dysfunction. Indeed, bad odors indicate the biological activity of organic matter degradation under anaerobic conditions (without oxygen) and the production of degraded organic compounds in reduced forms such as methane (CH_4), ammonia (NH_3) or hydrogen sulfide (H_2S). These elements in reduced form are no longer usable by plants and may even be toxic. In the presence of fresh organic matter, alcohols and aldehydes can also be produced, which are unfavorable to biological activity, seed germination, and plant growth. This phenomenon is encouraged by the burial of plants within a deep layer. When soil stinks, it is very often related to poor organic matter management, young organic matter in particular, and compaction. In this situation, the nose is an infallible analysis tool: anything that stinks is simply bad for the soil and for the plants.

13 • PHYSICAL FERTILITY ANALYSIS

The observation of the soil's structural condition is undoubtedly the most anticipated step for the farmer. Indeed, based on the diagnosis of physical fertility, the farmer can make immediate decisions regarding tillage or no-tillage for the next sowing, possibly undertake a deep restoration operation, and expect quick results.

13.1 Analyzing surface structural stability

Observing the soil surface provides information about its structural stability. The presence of a crust in the top few millimeters, or even the first centimeter, of the soil indicates a lack of structural stability. Every farmer knows how detrimental the formation of such a crust can be for crops, especially for their establishment. This phenomenon often causes the need to replant crops like beets, which have weak emergence power. Conversely, a surface with a crumbly appearance indicates a structure more conducive to water circulation, gas exchange, and the respiration of the soil, seed, and plant. A crumbly surface structure is also often a positive sign of surface biological activity, due to fungal activity with its mycelium and polysaccharide production, as well as microalgae, which have demonstrated significant capabilities in stabilizing surface conditions, even in arid environments (Isichei, 1990).

Slake soil test

Structural stability can be evaluated using the water glass test/slake soil test. The test involves placing a clod of soil on a coarse-mesh metal grid, immersing it in water, and measuring the time it takes for the clod to disintegrate.

The following stability scale can be used:

- 1, very low, 5 seconds.
- 2, low, 30 seconds.
- 3, medium, 5 minutes.
- 4, high, 30 minutes.
- 5, very high, 2 hours.

A standardized test called the Soil Stability Kit, developed by the USDA, is available from www.forestry-suppliers.com.

13.2 Compaction zones, plow pans, and hardpans

13.2.1 Compaction zones and plow pans

▪ Challenge: A major limiting factor

Paying closer attention to compacted zones in the soil is crucial when performing a soil profile analysis. These continuous zones (plow pans) or discontinuous zones (wheel compaction, for example) are always a major factor in the loss of efficiency of the cultivated system. A well-functioning soil is one that is connected from its surface to the bedrock by flows, transfers, and exchanges. In contrast, plow pans hinder the vertical functioning of the soil. They promote the formation of perched water tables, creating reductive hydromorphic conditions above them. The fertility of the compacted zone remains unutilized. The presence of these plow pans makes the cultivated plant less competitive compared to other weeds that are better equipped in terms of root systems to explore this compacted horizon. The presence of a compacted plow pan creates a "shadow" effect, significantly limiting the volume of soil explored by plants in the deeper horizons. Roots that bypass a compacted soil volume will continue to grow vertically due to geotropism (see Figure 12.1). All soil columns vertically above these compacted "plates," unexplored by roots, represent untapped potential, but also sources of leachable element loss since there are no roots present.

▪ Detection method

Plow pans never establish themselves permanently in the most superficial horizons. These are spared due to the intense biological activity present and exposure to climate variations such as dry/wet and cold/heat cycles. Generally, plow pans are located between 30 and 50 cm deep. Compacted zones are highlighted when refreshing the soil profile.

- **By the resistance to knife penetration:** The soil in compacted zones offers more resistance to the blade. These zones can also be identified by the presence of rust spots (signs of Fe(III) deposition, after migrating as Fe(II). These zones thus regularly show a different color from the rest of the soil and are frequently sites of detours or dead ends for anecic earthworms and roots.

▪ The penetrometer as a complement to the soil profile

Using the penetrometer alone quickly shows its limitations. Indeed, the resistance offered by the soil depends not only on its structure but also on its moisture conditions. It is very difficult to evaluate the contribution of each factor. Moreover, the presence of a compact horizon can resist the penetrometer while still having satisfactory porosity. This is often the case with old plow pans that have been reinvaded by microfauna, roots, and microbiology. You can have "aerated concrete": hard, but porous. Only after performing a soil profile can the results of a penetrometer be fully interpreted. However, it is an excellent tool to complement the profile in characterizing a phenomenon detected during the profile across an entire plot. But before extrapolating, it is essential to know exactly what is being measured.

- **By the differentiation in color:** It often happens, due to successive redox phenomena in the horizon immediately above the compacted zone, that the soil color is "washed out" of its pigments (iron and manganese), which accumulate lower down in the plow pan (see description of the phenomenon below). The use of a Munsell color chart can be very useful in this regard, providing more precise values than just adjectives (dark brown, brown/red, etc.), which are subjective.
- **By observation with a binocular magnifier:** To determine with certainty that the considered zone is not only hard but also lacking in porosity, it can be useful to examine a clod of soil with an optical device (×20 magnification). In the field of view of the device, one should be able to observe at least five or six biopores (round holes created by roots and maintained by microfauna). Sometimes, compact but porous zones are found! A functional soil should present between three and five pores per square centimeter of soil. When this figure is reached, the proportion of air in the soil is often 15 to 20%. The onset of anaerobiosis occurs as soon as a volume of one cubic centimeter of soil contains no porosity.

The horizon beneath the plow pan is often well oxygenated!

> It's paradoxical, but the soil underlying the compact horizon is almost always well oxygenated. It is indeed the site of oxidation reactions, visible notably by rust traces (iron oxide). The upper surface of the plow pan is quite impermeable, so water circulates very slowly, and gas diffusion is difficult within the massive blocks. However, the oxygen that diffuses through a few passages, cracks, or biopores crossing the plow pan can then spread throughout the entire underlying volume.

Tillage exacerbates soil compaction for physical and mechanical reasons. Indeed, clods are manufactured by humans using the same process used to make tiles and bricks: wetting, kneading, and then drying. When working with moist soil and allowing it to dry, the same phases are observed. Regarding deeper tillage, techniques from civil engineering are often employed: to prepare land for construction or highways, the natural soil structure is disrupted by using a ripper to break up porosity and disrupt binding agents (glomalin, mycelium, biopores, biological or chemical "cements"), and then compacting it. This increases its density and bearing capacity. If one pass is not sufficient, a second or even a third pass may be made to achieve the desired objective.

Is it always appropriate to talk about plow pans?

> Talking about a "plow pan" can be misleading because the semantics presuppose in the reasoning that the cause is known and singular. However, this is only partially true. The depth control of plowing is ensured by the part called the "heel," which impresses strong compression on the soil. Certainly, the plow is the tool that has been most widely used with this effect of creating a plow pan, but other tools can amplify or maintain it. Cultivators with tines (chisels), disc harrows equipped with plain disc blades and working with very tight angles, or power harrows can also create smoothing by the action of the working parts, which are often "dragging" (with a tight angle, 60 to 80 degrees in relation to the ground) and thus transfer the weight of the tool to the bottom of the tilled area. But the main subject is mainly this: soils no longer have the required resilience, the shape memory conferred by mycelium, root density, glomalin, etc. to regain their initial state after the passage of the tool. Physics is also dependent on biology.

13.2.2 Hardpans and iron pans

In connection with the emergence of Regenerative Agriculture and no-till practices, we have observed quite frequently, over the past five to ten years, deteriorating agronomic results among many farmers. Examination of the soil profile in these farms often shows the establishment of a compact layer that we can refer to as a "no-till hardpan," as its appearance is closely linked to no-tillage practices. These layers of compacted subsoil appear under temperate climates in sandy-siliceous, sandy-loamy, and loamy-sandy soils (we are referring here specifically to mineralogical siliceous sands, not granulometric sands that could be calcareous). They are established between 20 and 40 cm deep. They are identifiable by their highly compacted nature and differences in color between horizons: a lighter color, even "ashen" in the leached horizon, and sometimes darker shades stained with rust in the lower part of the compacted mass horizon.

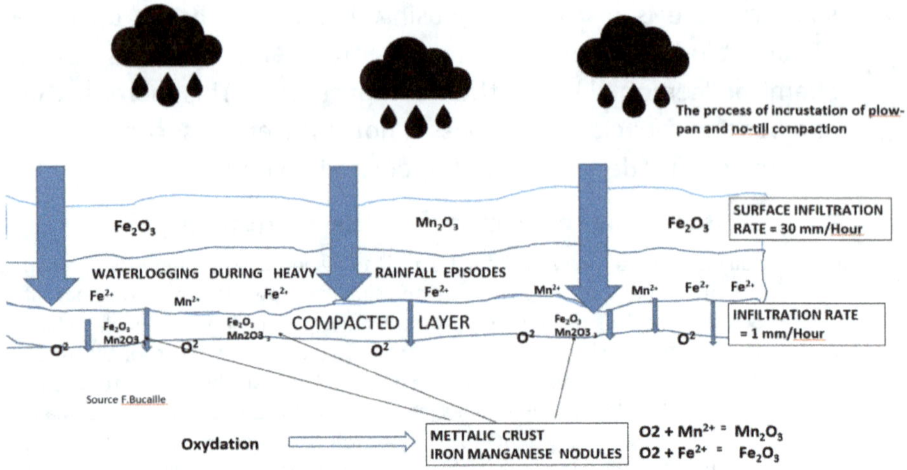

Figure 13.1 – The mechanism of hardpans hardening under no-tillage system

■ Onset mechanism

How do these compacted layers form? We have seen that cultivated soils have only two "tools" to ensure their restructuring and aeration. These are swelling clays and anecic earthworms, which create passages that allow water to infiltrate. However, sandy, loamy-sandy, and loamy soils contain neither. Until recently, these soils were worked with plowing. This plowing was generally far too deep, but at least it played a temporary role in aeration and allowed water to circulate at the bottom of the furrow. When these soils become compacted and collapse on themselves without self-restructuring mechanisms (due to the absence or insufficient quantity of clays and anecic worms that never favored these soils), the environment regularly becomes anaerobic and reductive during each major rainfall event. This is exacerbated by the fact that crops absorb much less water than the forest ecosystem from which most of our soils originate.

Reduced iron ions (FeII) are soluble in water and are transported to the deeper, less permeable layers of the soil. As they pass through these compacted layers, which are aerated at their lower part, the ions oxidize and precipitate in an insoluble form as Fe_2O_3, which is rust-colored. Iron is not the only element involved. Manganese cations are also implicated in the same process. As a result, within a few years, a compact layer forms, cemented with these metallic deposits, resembling podzolic phenomena. This mechanism operates as a vicious cycle. The more impermeable the compacted layer becomes, the more it retains perched water tables for extended periods. Consequently, the precipitation of metallic ions within the compacted layer is further amplified. Temporary anaerobiosis also impacts the evolution of organic matter and humus. In the presence of anaerobic bacteria, gases (NH_3, CH_4, CO_2...) and harmful substances such as alcohols and aldehydes (including formaldehyde) are produced, which are detrimental to microbiology. Similar to a slurry pit, there is a destruction of organic matter. The minerals, now separated from the organic matter, are prone to leaching, precipitating, sedimenting, and thus reinforcing the compacted layer. In a slurry pit, the sedimentation process generates hard, concrete-like precipitations through the same biochemical process.

When these phenomena are observed, it is urgent to break up the soil, which will immediately halt this mechanism and allow some of the deposited salts to dissolve and reach the deeper layers. The farmer's commitment to direct seeding and no-till farming should not prevent them from occasionally performing this beneficial operation.

■ Damage: ecological and economic losses, especially of nitrogen

The formation of a no-till compacted layer (hardpan) is accompanied by surface acidification. Water infiltration and recharge are poor, leading to high drought sensitivity. Nitrogen input requirements are high but leave very little residue. In this case, leaching is not the primary concern. Nitrogen needs are perceived as high because the frequent presence of perched water tables causes massive losses through gaseous emissions and denitrification (NO2). The plant's nitrogen uptake is poor because the nitrogen forms are not plant-assimilable. In these situations, it is crucial not to interpret the high nitrogen needs as a sign of humus stock replenishment, as is often assumed. This is not the case; the humus stock does not increase. The efficiency of the applied nitrogen is simply disastrous.

The formation of these no-till compacted layers in soils is far from marginal. In loamy and sandy-loam soils, we have observed that about one-third of the areas managed with no-till practices for five to ten years are beginning to be affected.

■ A phenomenon sometimes encountered on a large scale in nature

Generally, the natural ecosystem naturally opposes the separation between soil horizons. The deep rooting of large trees, in particular, ensures soil continuity and the recycling of minerals. However, some soils naturally exhibit this phenomenon of metallic hardpan in their climax state. These are called podzols. They form in natural ecosystems on rather acidic parent rocks and in the presence of climax tree vegetation, often conifers, which, through significant production of fulvic acids, cause mineral leaching. Additionally, this vegetation, which has relatively low and seasonal water needs, is not able to effectively counteract the leaching process. The Landes de Gascogne forest is well known for being the site of a specific case of podzolization, which causes the formation of a horizon of sand cemented by iron and manganese oxides, known as iron pan (alios layer).

13.2.3 Compaction: primarily a biological phenomenon

■ The compaction process

Soil compaction is the result of pressure exerted by agricultural machinery. However, the pressure from tires, and its consequences, only reveal previous dysfunctions, mainly of biological origin. These dysfunctions result from the lack of active roots producing root exudates and the absence of mycorrhizal fungi producing glomalin. In a typical undisturbed soil, fungi, especially mycorrhizae, continuously produce glomalin. This acts as a weak glue that maintains the cohesion of micro and macro-aggregates, thereby improving soil structure stability. Mycelium acts like tiny "elastic bands", providing micro-aggregates with mechanical properties of "shape memory", returning them to their original position after the passage of heavy machinery. This can be complemented by the effect of dense root systems and high organic matter content. Glomalin maintains cohesion and prevents crusting, erosion, and compaction; while mycelium offers resilience and the ability to restore the fine internal arrangement of aggregates. These mycelia, with diameters ranging from 2 to 10 microns, can penetrate the smallest interstices of a compacted volume and, like ivy between stones, separate compressed layers and reintroduce porosity. This effect is long-lasting because glomalin, a glycoprotein produced by mycorrhizal fungi, is highly stable over time, with half-lives ranging from 15 to 40 years. While this is advantageous, its longevity has led to a belated recognition of the negative effects of degrading agricultural practices. Their positive effects were still noticeable even when the microbiological agents responsible for them had largely disappeared.

■ **Bacterial glues**

Disturbed soils contain fewer fungi and more bacteria. Bacteria exhibit very different individual and collective behaviors. They live in colonies and exude, mostly polysaccharides, bacterial gums that have the significant disadvantage of strongly binding aggregates and micro-aggregates together when compressed, without hope of spontaneous restructuring. Heavy agricultural equipment compresses micro-aggregates, chemically and physically binding them together, thus compacting the soil permanently. We have diagnosed severe compaction in situations in Brazil that seem virtuous, namely alternating periods of cultivation for five years with a one-year ley. It turns out that these systems generate the most severe compacting conditions. The explanation is that cattle impose significant physical constraints (the pressure per square centimeter of the animals' hooves - up to 15 Kgs/cm^2 -is much higher than that imposed by an agricultural tire - 0.8 Kg to 3 Kg/cm^2), without the ley having had time to re-establish, in just one year, an active microbiology with significant fungal presence and root mass. The good structural states of natural grasslands are therefore not due to fewer agricultural machines but primarily to a better ability to withstand mechanical stresses. Machinery is not the cause of all our troubles; it merely highlights that our soils are ready to collapse due to the absence of the assembly parts that previously held the structure in place.

Soil compaction is above all the expression of a biological problem linked to the reduced production of polysaccharides and glomalin in the soil. Soil compaction is due to a lack of living roots and fungal mycelia in the soil.

Decompaction: misleading semantics

The term "decompaction," by its very nature, suggests that a mechanical operation alone can repair structural damage caused by long-standing practices or unfortunate agricultural interventions. However, this is not the case. Decompaction is nothing more than cracking a horizon that has become compacted. The result of the operation will be the creation of blocks, measured in decimeters, which still have not regained the lost porosity. While we may have restored water infiltration capacity and created passages for roots, these blocks will not immediately be suitable for hosting biological activity, and they will not initially contribute to plant nutrition. This zone will only host transport roots, not assimilation roots. To avoid misunderstanding, let's call this operation what it is: cracking.

13.3 Porosity, the soil's respiratory system

Porosities form the respiratory pathway of the soil: macropores form the tracheae, and micropores form the pulmonary alveoli. Establishing a high level of porosity, corresponding to a 25% air volume in the soil, is essential to ensure soil respiration and limit the occurrence of anaerobic conditions.

Biopores

Biopores are a physical characteristic of soil that is closely connected to biological factors. They largely contribute to the presence of soil microporosity. Their origin is biological, linked to the activity of earthworms, as well as microfauna such as enchytraeid worms and roots. The abundant presence of these small channels dug into the soil is a sign of a surface structure favorable to cultivated ecosystems. Biopores can be detected with the naked eye, but observation under a binocular microscope allows for a better understanding of their abundance and functional effectiveness.

13.4 Clays: the best ones always leave first

13.4.1 Stability of clays in the soil

The challenge is to counteract the accelerated aging of soils, namely the loss of the best clays. As we have seen, the binding between humus molecules and clays, facilitated by cations, causes them to flocculate. This "protects" them, with one from mineralization and the other from leaching. When the humus content decreases, the direct consequence of breaking this beneficial bond is that some clays become "orphans". These clays are then at risk of leaching, being carried by water to deeper horizons, and accumulating towards a new horizon called the illuviation horizon. The effects are as follows: the surface horizon becomes increasingly more prone to crusting, the deeper horizon becomes more impermeable, leading to longer drying times. Finally, the cation exchange capacity (CEC) of the surface tends to decrease. It is obviously the clays with the largest internal surfaces that are subject to this transfer since they have powerful electrical charges, giving them a superior capacity to suspend in water as a colloid

Note: this phenomenon of surface clay losses is irreversible.

13.4.2 Deep soil structure

The primary quality of a living soil is to have efficient pathways for air and water movement between the surface and the bottom of the agricultural soil (up to two or three meters deep for the best soils). Anything horizontal within the depth of soil, with more or less deficient porosity, is detrimental to the performance of the soil-plant system.

Heavy, clayey soils may exhibit characteristics in depth related to the structural activity of clays. These are the result of cracks that always occur in the same places, in response to various climatic hazards (primarily alternating moisture and drought). The deep structure of the soil is an inheritance that cannot be agronomically intervened upon, except through considerable means such as drainage or deep subsoiling (for layers of alios). Analyzing these structures is important, not necessarily to modify them, but to understand the agronomic potential of the soil and guide practices. These characteristic structures are highlighted during knife-profile soil refreshing. We will not delve into the detailed classifications of pedologists, which are well documented in specialized literature, but rather focus on the four main types of agronomic situations commonly found in cultivated soils and their implications for farmers.

- **Columnar and prismatic structures.** Clays are organized into "columns" through shrinkage phenomena. These vertical columns have a polyhedral shape when viewed from above, at a decimeter scale, and can range in height from a few decimeters to two meters. These soils can accept significant amounts of water when precipitation resumes in the fall (when there is drainage, the drains react within a few hours, posing a risk of leaching). As soon as the clays swell and close again, these soils become highly impermeable and can once again become waterlogged in case of excessive rainfall. During rainy years, clay restructuring through cracking does not occur, making autumn sowings much more difficult. In all soils of this type (clays with shrinkage cracks), it is strongly advised not to work mechanically during dry periods because the operation will result in fine surface elements falling deep into the soil. This results in surface fertility loss and deep horizons compressed during rehydration, making them even more impermeable. This type of structure is easy to identify by the presence of large vertical crack planes at the decimeter scale, identifiable with a knife.

- **Massive structures.** This is a structure that does not exhibit distinct aggregates. Many deep clays exhibit this appearance, often characterized by blue, gray, or green colors. They are generally subject to shrink/swell phenomena as well. Successive shrinkage events over time can create large volumes, sometimes at the meter scale, with characteristic oblique slip planes (called slickensides). These highly cohesive clays are not conducive to dense and evenly distributed root systems. Additionally, there are never any anecic earthworms in these conditions, so there is no hope of introducing them through even the most virtuous practices. These soils can experience waterlogging on the surface. Some sandy soils can also evolve into massive structures (without the typical shrinkage phenomena of clays). This evolution can occur naturally, such as through podzolization, which leads to the fusion of sand grains with iron and manganese oxides acting as cement. Farmers can also be a cause by planting resinous tree species for example, in which case acidification is the cause, or by planting crops incapable of absorbing the average rainfall of the region (grass species are adapted to steppe conditions with 400 mm/year compared to 800 mm/year and more for tree species). In the latter situation, temporary waterlogging due to the mismatch between vegetation needs and the soil's capacity to drain excess water will quickly undermine the initial favorable structure of these soils. We have seen soils develop this mechanism under grassland, then resolve it and manage to reinvest the deeper layers, beyond the iron pan that formed, by reintroducing apple orchards. The latter are better suited to the climax of our temperate regions.

- **Particulate/unorganized structures**. There is no visible organization of aggregates. There is no particular orientation of the coarse constituent elements, which appear individualized and autonomous. These may be sandy or silty soils (greater than 20 microns). Generally, root systems are both dense and well distributed. Since these soils are generally not very clayey, they do not spontaneously restructure and are therefore fragile. Sandy and siliceous soils, as well as sandy loams of this type, do not host deep anecic populations, further increasing their susceptibility to compaction. Soils predominantly composed of silt of this type can evolve under the impact of agricultural operations (plowing or direct seeding) towards a structural organization into "plates" (see below), while sandy soils can evolve towards massive structures (see above).

- **Plate-like structure**. Modeled after shale or slate, this soil architecture can be inherited geologically or created by human action. When of natural origin, it is often the case that the orientation of the "slate sheets" is oblique. When of anthropogenic origin, within the top 50 centimeters, they are perfectly horizontal. These structures have implications for all indicators: slowing down water circulation, root bypasses, and hindrance or disappearance of anecic fauna. This physical obstacle can be further amplified in its impact by mineral precipitation due to biological phenomena described elsewhere in the book. When this phenomenon occurs, mechanical cracking (subsoiling) is often unavoidable, whether there are cover crops or not.

13.5 Color: a reflection of soil evolution and functioning

The different colors of soil horizons can be determined using the Munsell Soil Color Chart reference. Color provides information about the level of iron oxidation (warm hues like reds, browns, oranges indicate good oxygenation), the presence of gley (waterlogged soils with blue/green colors), as well as leaching and eluviation phenomena of soil pigments such as iron, manganese, and humic substances. The benefit of using the Munsell chart is to allow for comparison of two pedological situations over time or space. Since this chart is used worldwide, it facilitates communication among agronomists.

14 • CHEMICAL FERTILITY ANALYSIS

Chemical phenomena are both the cause and consequence of the biological and physical processes previously discussed. However, there are some purely chemical indicators that are interesting to measure on-site during soil profile diagnostics. These procedures, however rudimentary compared to laboratory procedures, still have practical value, particularly useful for decision-makers such as technicians or farmers. Their reliability lies in their repeatability and the ability to locate them finely within the profile height, allowing them to be linked to ongoing developments.

14.1 pH

14.1.1 Challenge

Surface acidification is common and exacerbated by certain agricultural practices such as conservation tillage systems. One of the main causes of surface acidification in cultivated soils is the leaching of cations such as calcium, magnesium, potassium, etc., towards the deeper horizons. This phenomenon can be amplified by the leaching process of nitrates or clay leaching and the release of hydronium ions. pH measurement is a good indicator for detecting these mechanisms.

14.1.2 Method & analysis

We propose here a method for analyzing pH to identify a vertical acidity gradient in the soil. To study pH, we recommend using a pH meter of the Hellige type that utilizes a liquid reagent operating over a wide pH range (from 4 to 9). While the precision may not be very high, this type of measurement is very reliable and quick to perform in the field. We recommend conducting at least three pH measurements on a soil profile. One in the surface horizon from 0 to 5 cm, the second just at the upper level of the compacted horizon, and the third at the deepest level of rooting, or directly on the bedrock if possible. Often, a gradient is observed with a pH evolution in depth.

- **The first surface measurement**, from 0 to -5 cm, will allow for the identification of any surface acidifications, which can be very pronounced, especially in no-till systems or perennial plants. Irrigations amplify the phenomenon. A soil analysis based on a conventionally sampled 20 cm depth will not reveal this. Even soils saturated with calcium are not immune to this trend.

- **The second measurement** taken in the upper face of the plow pan/no-till hardpan (approximately 30 cm deep) will confirm or refute the leaching of bases. A one-point difference between the second measurement and the first is already highly significant. This phenomenon is often the manifestation of a lack of fungi (see the section on pseudomycelia). We have already identified differences of five pH points between -5 and -20 cm! (pH 4 on the surface and pH 9 at 20 cm depth). All analytical indicators were green, including pH at 6.8 (sample taken at a depth of 20 centimeters). Fruit trees were dying despite this favorable average!
- **The final measurement** taken at the bottom of the pit provides an idea of the nature, basic or acidic, of the deep rock upon which the surface soil rests, whether it originated from it or not. If we have deep-rooted plants, this can help manage surface fertilizations by focusing only on the early needs of the crop, which is especially valid for perennial crops (vineyards, fruit trees...)

14.2 Presence of calcium carbonates: hydrochloric acid test

14.2.1 Challenge

Detecting calcium in carbonate form provides immediate practical information by identifying one of the possible causes of antagonism between elements: iron chlorosis, phosphorus uptake, etc. This test will guide the farmer towards fertilization methods suitable for soils with a high pH and almost no H^+ ions. Decarbonation is often linked to soil aging, which can be accelerated by acid rain or the use of acidifying fertilizers. Soils that are largely decarbonated show less pronounced blockages concerning iron, phosphorus, and other micronutrients.

14.2.2 Method

Use diluted hydrochloric acid at a 1:10 ratio (pour the acid into the water, not the other way around) and pour a few drops onto the soil block to be tested. Observe whether effervescence occurs throughout the volume or on localized particles. If effervescence is only localized to coarse sands and gravel, calcium antagonism likely won't exert as strong an effect on plants because entire volumes of soil will offer a favorable space for soil-root exchanges. The intensity of the reaction can be classified as follows.

Reaction to HCl	Calcium Carbonate Content	Part of $CaCO_3$
Absent. Neither visible nor audible	None	< 0.5 %
Not visible, but audible	Presence	< 1 %
Detectable	Traces	< 2 %
Weak	Weak	2 to 10 %
Medium	Medium	10 to 25 %
Strong	Strong	25 to 55 %
Very strong	Very strong	> 55 %

Table 14.1 – Analysis of calcium carbonate content based on hydrochloric acid test results

This method allows for the identification of a potential gradient throughout the soil profile. As described, it will only react with calcium carbonates. If we want to detect the presence of dolomitic limestone ($CaCO_3$ associated with $MgCO_3$), in a second phase, we need to heat a small sample of soil soaked in hydrochloric acid with a lighter placed on the tip of a knife (without heating up to the boiling temperature). If there is effervescence, then we can conclude that there are also magnesium carbonates.

15 • HYDROLOGICAL FUNCTIONING ANALYSIS

Soil hydraulics are complex, and depend on soil structure, porosity, biopores, texture, rooting depth, calcium/magnesium ratios and humus content. Good water fertility is defined as the capacity to absorb excess water without erosion or hydromorphy (asphyxia), to store most of this water deep down and then release it to plants when rainfall is insufficient. Developing and maintaining this storage capacity is essential if we are to reproduce the natural climax model in agrosystems (see chapter on climax).

15.1 Identifying a soil's water absorption capacity

A well-functioning soil is generally capable of absorbing 35 mm of water in less than five minutes in the surface horizon. Measuring this water infiltration rate is feasible across different soil horizons. To conduct this measurement, simply dig a step in each of the horizons to be studied, place a simple tube with a diameter of 15 cm, and pour the equivalent of a 35 mm water column into it. The measurement can also be done more precisely using a double-ring infiltrometer, such as the Guelph permeameter.

These infiltration measurements should ideally be conducted on a dried soil. The infiltration duration for most surface horizons is one to three minutes. This time can be much longer in compacted horizons and last up to 24 hours! A five-minute flow duration in the compacted horizon is considered very good. Beyond 15 minutes, the situation is deemed at risk from a waterlogging perspective. These water stagnation measurements inform us about the sensitivity of soils to waterlogging. The difference in infiltration time between the surface horizon and the compacted horizon (-30 cm) is crucial for identifying the risks of perched water table formation during intense rainfall events (from a difference of 15 minutes, soils can be considered at risk). However, it should be noted that temporary waterlogging does not necessarily indicate the systematic establishment of anaerobic soil functioning (without oxygen). Indeed, there are aquifers, especially alluvial aquifers, rich in dissolved oxygen. Rains also bring dissolved oxygen, so reduction only occurs a few days after the end of the rainfall event (Baize et al., 2014).

15.2 Detecting signs of waterlogging or anoxia

15.2.1 Challenge

Hydromorphy in a soil refers, as its name suggests, to the set of morphological signs related to waterlogging over a sufficiently prolonged period to create so-called reducing conditions. Trapped or dissolved oxygen is depleted by the activity of aerobic microorganisms within a few hours or days. Oxygen scarcity, or even anoxia, paves the way for the development of anaerobic microorganisms. These microorganisms respire through the reduction of various compounds, including iron and manganese. The alternation of these periods of waterlogging with periods of drying (reoxygenation) is manifested in the soil by various traces.

Hydromorphic soils are those that regularly experience periods of waterlogging for varying lengths of time throughout the year. Hydromorphy, and the associated anoxia, negatively influence the mineral balance of the soil and the entire soil biocenosis in cultivated areas. Sometimes, anoxia is not directly caused by waterlogging. It can also originate from the excessive presence of fresh organic matter in depth. Indeed, as we have seen before, the degradation of fresh organic matter in depth often leads to a biological oxygen demand (BOD) greater than the oxygen availability in that part of the soil. Traces of this "biological" anoxia can be differentiated by odor, but also by the bluish color imparted to the soil by reduced iron ions.

15.2.2 Method and analysis

■ Traces of iron and manganese

The conditions of anoxia are frequently detectable by bluish or greenish traces of reduced iron or, more typically, by the formation of ferromanganese nodules. Almost always, a hydromorphic soil will exhibit these types of characteristics. Here's how these signs of hydromorphy form. In the absence of oxygen (O_2), bacteria (*Acinetobacter, bacillus, pseudomonas*...) seek "electron acceptors" other than the oxygen in the air, which is lacking. They find them, notably, in the oxygen atoms associated with iron and manganese oxides. The metallic elements separate from their oxygen atoms and then concentrate as reduced, cations soluble in water, which they follow the path of. When the water recedes, air replaces the water in the soil pores. Iron and manganese reoxidize by donating their electrons to oxygen. They become insoluble again and deposit as metallic traces or as nodules of variable size, ranging from a tenth to a few millimeters in thickness. These black ferromanganese nodules always form in the same places. Indeed, water flows and paths exhibit great consistency in soils.

■ Migration distance

The migration distance of iron and manganese also varies from a few centimeters to several meters. The distance between the nodules and the soil surface provides a good indicator of the duration of hydromorphy during the year. When nodules are located within 25 centimeters, it indicates mild and temporary hydromorphy. Nodules found at depths between 80 cm and 1.8 m indicate periods of hydromorphy lasting several months during the year.

■ Other traces

Other traces are characteristic of hydromorphy. This may include partial or total depigmentation of the soil associated with the disappearance of iron (referred to as albic soil matrix). Reductic horizons can also be identified by a greenish to bluish coloration due to the presence of ferrous ions (reduced). This indicates recent or ongoing hydromorphy. As soon as the soil reoxygenates, the coloration fades due to iron reoxidation. This reductic feature is difficult to observe in very clayey dark horizons.

Hydromorphy: the usual is not normal

In some regions of northern France or in Belgium, soils often exhibit a blue horizon. This occurs so commonly in the region that farmers consider it completely normal. It may be usual, but it's not normal. In this case, agricultural practices are to blame. This blue layer, indicative of negative redox conditions in the soil with the presence of reduced iron, is linked to the burial of organic matter in depth, compaction, and the loss of porosity in soils made particularly sensitive to anoxia.

■ A homemade hydromorphy test

The presence of reduced iron (associated with hydromorphy) can be easily detected in the field using a homemade reagent made from 2% potassium ferricyanide solution supplemented with 2% hydrochloric acid. This solution is applied to the suspected soil clods using a dropper. This helps determine if a gray horizon has this color due to the loss of its pigments, including iron, or if iron is present in the form of Fe(II) ions. If the reagent turns intense blue, it indicates the presence of ferrous ions (reduced Fe^{2+}) indicating that the analyzed soil section is affected by hydromorphy. Otherwise, the reagent remains yellow. However, the results of this test require interpretation. Indeed, intense biological activity in surface horizons with a high biological oxygen demand can also lead to a negative redox and the presence of reduced iron. The abundant presence of life signals, porosities, and tunnels allows us to identify cases of "false positives" for hydromorphy. Similarly, in compacted horizons, the reagent often shows a yellow color. As we have already explained, the lower side of compacted horizons is paradoxically often well-oxygenated but also very low in fresh organic matter, resulting in a very low biological oxygen demand.

Ferrous and ferric iron

> Iron exists in the soil in two different states. In a reducing environment, it accepts electrons from redox reactions mainly associated with the respiration of anaerobic bacteria. Iron is then present in the reduced form of ferrous ions ($Fe2^{+}$)which combine to form compounds such as ferrous oxide FeO (bluish in color), ferrous hydroxide $Fe(OH)_2$, ferrous carbonate $Fe(CO_3H)_2$ and ferrous sulfate FeS. All these forms are highly soluble in water. When solubilized ferrous iron encounters oxidizing conditions in the presence of oxygen, it releases its electrons. The iron thus takes its oxidized form as ferric ions (Fe^{3+}). This insoluble form in water precipitates as ferric oxide (Fe_2O_3) or ferric hydroxide $Fe(OH)_3$.

■ Special cases of iron-free soils

The characterization and detection methods of hydromorphism described earlier are based on observing the behavior of iron in the soil. This principle is also recognized by French legislation on wetlands. However, the regulation specifies that "in certain specific contexts, prolonged excess water does not result in the usual easily recognizable traits of hydromorphism." This is particularly the case for fluviosols derived from materials very poor in iron, humic and humoduric podzols, sandy quartz-rich geological materials, or soils completely leached of iron by numerous cycles of redox reactions (Baize & Ducommun, 2014). In these particular cases, on-site observations by the farmer of the water regime are crucial.

15.3 Evaluation of water fertility

Soil has the capacity to store excess water and release it during periods of drought. This ability to absorb water without erosion or horizontal run-off, and then return it to the plants, is one of the fundamental criteria of soil fertility. An ideal soil should contain 25% of its volume in water at the drainage point. Water fertility is an element that man can influence directly, through irrigation or drainage for example, and indirectly, by optimizing soil functions such as porosity and organic matter content.

15.3.1 Field capacity (= drained upper limit: DUL)

The type of soil greatly influences its ability to sustainably store water. Sands infiltrate water very quickly but are not very good at storing it. Clays, on the other hand, infiltrate water very slowly but can store it for a long time. To measure the soil's water retention properties, it is common practice to measure or calculate the field capacity. Field capacity is the soil's ability to receive water until it reaches the point of drainage. In other words, it is the maximum amount of water the soil can hold until it begins to percolate, expressed as the quantity of water per quantity of dried soil. It can also be measured by dehydrating a sample taken at field capacity.

15.3.2 Readily Available Water (RAW)

Readily Available Water (RAW or hydric comfort) corresponds to the quantity of water readily available in the soil for the plant, when the soil is dry (at field capacity). It is expressed in millimeters and averages 10 mm of water per decimeter of soil. This indicator is therefore easily comparable with rainfall or irrigation, which are also expressed in water depth.

By determining horizon textures, thickness and rooting depth, it is possible to calculate RAW. For this purpose, there are abacuses that give each soil type an RAW in millimeters for each centimeter of soil thickness. RAW is a good indicator of water fertility. Soils with a high RAW, such as deep, well-structured silts, ensure superior water comfort for plants. Plantations will be more resilient, particularly in the event of poor rooting.

		Soil texture	RAW (mm/10 cm soil)
1	Si	Silt	1,80
2	SiL	Silty Loam	1,80
3	L	Loam	1,70
4	SL	Sandy Loam	0,90
5	LS	Loamy Sand	0,70
6	S	Sand	0,50
7	SCL	Sandy Clayey Loam	1,70
8	SC	Sandy Clay	1,70
9	CL	Clayey Loam	1,70
10	SiCL	Silty Clayey Loam	1,80
11	SiC	Silty Clay	1,85
12	C	Clay	1,85
13	HC	Heavy Clay	1,90

A capillary continuum may be additional to the RAW

The estimation of Readily Available Water (RAW) is sometimes underestimated because it does not account for water flux by capillarity. This is particularly relevant for very good loamy soils or calcareous subsoils, where the RAW can be corrected by several additional tens of millimeters of water height. This capillary phenomenon varies according to soil types and their operational state. The capillarity continuum can extend over a meter of soil height in very good loams, but it is limited to a few centimeters in sandy soils.

These values of water transfer, coming from the underlying horizons at the root limit of crops, can represent 100 mm for loams, 50 mm for heavy clays, and 5 mm for sands. Soil continuity breaks, as well as compacted zones, also diminish the effectiveness of capillarity, thereby reducing water fertility.

15.3.3 Water fertility and organic matter

It is commonly said that soil humus can hold water up to ten times its own weight! This topic is extensively documented. However, one much lesser-known aspect is that **organic matter and humus act as water savers for plants** for two main reasons, which go beyond their physical water storage capacity.

1. Humus is indeed a condensed mixture of plant tissues and living matter. Consequently, its composition closely resembles the tissues from which it originates. When there is mineralization and release of its stored reserves, the assimilable nutrients released into the soil solution are present in exactly the proportions required by most plants. Plants thus have access to the perfect menu. They have needs that must be met for each element within ranges acceptable to their own physiology. If a mineral imbalance occurs, the plant will continue to transpire in hopes of accessing the deficient mineral, without necessarily manufacturing and accumulating dry matter optimally. It will resume full production only when its needs for the deficient element have been satisfied. During this time, there will have been a loss of yield. The water use efficiency will have suffered. Most plants require between 250 and 750 liters of water per kilogram of dry matter produced. Optimizing plant nutrition via the soil and humus allows us to stay in the lower levels of this range. The optimization of nutrition from this perspective will be revisited in Chapter 17.
2. When humus mineralizes, it releases minerals and inorganic salts, as well as much more complex molecules such as amines, vitamins, and saccharides, which are all building blocks of life. A plant that absorbs macromolecules that it will reuse as they are will benefit from energy, growth, and water consumption. Some of these organic molecules,

especially humic acids, also have a biostimulant effect on soil biological activity, roots, and the entire plant.

Soil respiration and condensation: a lesser-known phenomenon

The soil undergoes a marked respiration process characterized by exhalation during the day and inhalation during the night. The origin of this phenomenon lies in temperature differences and the expansion and contraction of the gaseous phase on a circadian rhythm (day/night). The moisture in this nocturnal "breathed" air can condense on the walls of micropores, contributing modestly but regularly to improving its water fertility. The soil's ability to condense water from the air it breathes is largely dependent on its porosity and microbial life. Hence, the importance of high microporosity and the absence of surface sealing crusts. The adage "hoeing is worth two waterings" can contribute to this phenomenon when it helps break up a sealing crust. The soil's ability to "breathe" can be enhanced by the use of fertilizers and amendments (see Chapter 20).

16 • TESTS TO ASSESS A LIVING SOIL

Tests, known as biotests, seek input from plants, macrofauna, microbiological flora, and inherently incorporate a significantly larger number of parameters. This methodology does not exclude the use of more conventional chemical analyses. Each serves different purposes: Laboratory chemical analysis exclusively measures what it is designed to detect and can identify potential toxic or beneficial molecules. Living organisms, through their growth and vigor, synthesize all environmental factors, "remember everything," and even include those that are subtle, cumulative ("cocktail effect"), or unknown. There are numerous possible biotests. We will describe some useful, simple, cost-effective tests that allow farmers or their technicians to better manage fertilization and plant protection practices.

16.1 Garden cress germination test

Operating principle: Garden cress is easy to find in garden centers. It germinates very quickly. This plant is sensitive to the presence of anti-germination compounds, allelopathic substances, and other growth inhibitors or phytotoxic substances in the soil. When these factors are present, the germination ability of the cress is affected, and the developing plants exhibit irregular growth, appearing stunted and displaying yellow or brown colors.

Objective and usefulness: This test is primarily intended for organic materials. Garden cress can be used as an excellent bioindicator for the presence or absence of phytotoxic or phytorepressive factors in soil or organic amendments. To such an extent that a standardized method (NFU44-165 of December 2015 and AFNOR XP U 44-165) for characterizing composted raw materials has been developed based on the germination abilities of cress. This test is the most effective technique for detecting the presence of allelopathic substances in the soil, analyzing questionable compost, pruned wood, and ramial chipped wood (RCW), and ensuring they do not exceed the soil's neutralization capacity. It is more important to know this characteristic than the levels of N, P, and K, or even the C/N ratio. Indeed, the latter has only a distant correlation with the maturity degree of organic materials. A material can have a low C/N ratio (10) and still not be mature, with a high biological oxygen demand (BOD), such as slurry or materials rich in sugars, like green manure.

How it works: We recommend conducting the garden cress test on a substrate composed of 50% of the organic material and 50% of the soil intended to receive it, and comparing it to a control without compost in containers that can hold 5 liters. Seedling counts and the evaluation of the morphological state of the emerged plants should be done 3 days and 7 days after sowing. If there is more than a 20% difference between the control and the mixture, it will be advantageous and cost-effective not to spread the organic material, even if it is free. Thus, a product with favorable chemical analyses can prove disastrous in terms of soil amendment and vitality.

Advantages: This method is the only one that can detect anomalies that do not appear in standard analyses. These often include specific toxicities that may be added to the target soil (which itself may already contain heavy metals at levels close to phytotoxicity). Garden cress reacts to the presence of terpenes, polyphenols, coumarin (a substance meant to protect plant tissues from parasitic attacks and found in large quantities in cereals straw), heavy metals, and more. However, a simple mineral deficiency in the soil or nitrogen starvation does not have a toxic nature and is not detected by the garden cress test. This test definitively differentiates between true "nitrogen starvation" and allelopathy caused by immature plant residues (straw, RCW).

16.2 Herbicide toxicity test on oats and sunflowers

Operating principle: Certain plants have specific characteristics (physiology, rooting, etc.) that make them more sensitive to the action of herbicides (sunflower). Sometimes, their reduced economic importance has not driven the development of molecules or the improvement of genetics for tolerance to existing molecules (oat).

Objective and usefulness: Due to their significant sensitivity to many molecules, oat (a grass) and sunflower (a dicotyledon) are particularly well-suited to revealing residues that may be detrimental to a crop. Their growth will be visibly affected at levels lower than those that would express toxicity in the intended crop. This test is very useful in intensive systems.

How it works: Fill four containers with a capacity of approximately 5 liters with soil from the surface layer (the top 10 centimeters corresponding to the seedbed). Two containers should contain the suspect soil and two should contain uncontaminated control soil. Sow oats and sunflowers in both the control and suspect soils, using 25 seeds for each condition.

Water the plants, wait 21 days, then extract the entire plants, dry them, and weigh them. Any difference greater than 10% will be significant.
Advantages: Simple, inexpensive, and relatively quick (3 weeks).

16.3 Earthworm toxicity test

Operating principle: Earthworms play a crucial role in soil formation (pedogenesis) and in maintaining soil functionality. Their digestive system relies on enzymes and microorganisms, making them a living indicator of the soil's digestive capacity. Earthworms are particularly sensitive to all substances that are aggressive to their mucous membranes, such as ammonia, alcohols, pesticides, insecticides, essential oils, plant extracts, molluscicides, and fertilizers.
Objective and usefulness: This method evaluates the biological value of organic amendments, composts, and ramial chipped wood (RCW) for macrofauna, as well as the toxicity of soil (e.g., copper), fertilizers, or amendments.
How it works: All you need to do is buy 4 boxes of earthworms (easy to find on the internet from vermicompost producers), regardless of species. On receipt, simply remove the earthworms from their boxes, wash them gently in distilled water, without drowning them. Then weigh them and return them to their original boxes. Add 30 milliliters (adjust if necessary) of water to which fertilizer or active ingredient has been added, according to the following calculation:
Solution preparation: We'll consider the quantity spread per hectare: e.g. 250 Kgs KCl (potassium chloride). We'll take a typical amount of rainfall that would allow any product applied to infiltrate the soil's superficial horizon: 25 mm (i.e. 250,000 litres/Ha). Or 1g of KCL diluted in 1 liter of water (250 kg /250,000 l). Incorporate 30 millilitres of this solution into a can. It's a good idea to also test a double dose, i.e. 2g/litre (i.e. 500 Kgs KCl/Ha) and put 30 ml in the second can. To test crop protection products at lower volumes/Ha, the dilution should be as follows: 1 drop of the crop protection speciality/5 liters of water for the equivalent of a 1l/ha crop protection product dose. Then add 30 ml of this mixture to the first can of earthworms. For the double dose: 2 drops/5 liters, then add 30 ml to the second box. The last two boxes will receive the same quantity of distilled

water as the first two. Add half a teaspoon of cornmeal to each can to feed the earthworms. Place the boxes in a temperate location.
Wait one to four weeks. Observe their condition, re-sample, wash with distilled water and weigh. If the product tested is neutral, the weight will be the same as in the control boxes. If, on the other hand, there is an unfavorable effect, the gain will be less, or even a weight loss will be observed. This test can also be used for agricultural soils. Simply replace the original medium with soil, after washing and weighing the earthworms. Then add half a spoonful of cornmeal. A soil + fertilizer combination can also be considered.
Advantages: This test provides an overall assessment of soil vitality or the impact of a fertilizer or pesticide on a vital soil digestion process.

16.4 Soil or growing medium mycorrhization test

Operating principle: Mycorrhizae are decisive auxiliaries in the nutrition of many plants and in carbon storage. They can be sensitive to a number of compounds: phosphorus, heavy metals, terpenes, fungicides, etc.
Objective and usefulness: The test will involve testing the ability of a soil or substrate to allow mycorrhization of a typical plant: the highly mycorrhizogenic white clover (Trifolium pratense). The aim is to validate that the physico-chemical characteristics of the soil are not limiting for crops dependent on mycorrhization.
How it works: The plant used, clover, will be planted in 20 pots of around 250 ml. 10 will be inoculated (Novozyme or Bayer) and 10 will serve as controls on soil sterilized at 120°C for 20 minutes. The experiment will last 5 to 6 weeks. At the end of the experiment, the vegetative state will be completely different if mycorrhization has taken place. Mycorrhizae can also be directly observed by treating the roots with 10% caustic potash (KOH) for 1.5 hours and staining with 0.5% Typan Blue for 20 minutes.
Advantages: This test makes it possible to assess a particular and essential fraction of soil microbiology.
Disadvantages: Time-consuming. Direct observation of the vegetative state, or even weighing the biomass of plants, is simple, but if you use a binocular telescope, it requires a bit of know-how in preparation and interpretation.

16.5 Impact test on general or targeted soil microbiology

Operating principle: Microorganisms are sensitive to certain organic and inorganic molecules, both natural and synthetic.

How it works: Check the impact of the products used, fertilizers, phytosanitary products and natural preparations on soil microbiology. This allows you to visualize the reality of biology in agriculture, and to integrate this information into the characteristics of the inputs used.

How it works: Obtain Petri dishes (many suppliers: ATL, DIFCO, BIOKAR, HACH, VWR, ...). An agar-based nutrient medium is used. Three types of inoculant can be used. In general, the inoculant is obtained from a filtrate obtained from a soil sample, and sprayed onto a Petri dish containing a PDA (***P**otato **D**extrose **A**gar)* medium. Two families of soil microorganisms can be used: *Azotobacter chroococcum*, a nitrogen fixer, and *Bacillus mucilaginosus,* a PGPR (Plant Growth Promoting Rhizobacteria) bacterium. Soak filter paper slices in the solutions prepared as described above in the "Earthworm toxicity test". Place filter paper discs in the middle of Petri dishes. Incubate between 72 and 96 hours. If the solutions are harmful to the development of microorganisms, you'll see a clear, more or less marked area around the soaked paper.

Advantages: A simple, inexpensive test that provides invaluable information to guide the choice of products and technical itineraries, and to help achieve the main objective, which is soil vitality, and therefore economic performance.

16.6 Bacteria/fungi balance test

Operating principle: Bacteria and fungal mycelia have a different appearance and can be differentiated visually (by dyeing if necessary). A field kit combined with an online application enables you to obtain the information you need from your cell phone in less than an hour.

Objective and usefulness: Measure total bacterial biomass and identify bacterial/fungal balances in relation to agricultural practices. When it comes to microbiology, it's the harmony between different communities that's more important than the overall volume of activity. In cases where fungal families are under-represented (the most frequent case), corrective measures can be immediately envisaged and implemented: Ca/Mg amendments, addition of woody organic matter, etc. An American company, microBIOMETER®, is specialized in this technology: https://microbiometer.com/

How it works: 1) Using a multicomponent sample taken from the field to be studied, sieve the soil and place a sufficient quantity in a plastic bag. Remove coarse elements and any insufficiently decomposed organic matter. 2) Add the extraction powder to the test tube, then the water, and finally 0.5 ml of soil 3) Whisk everything up and leave to stand 3) Take the liquid and place 3 drops on the test chart provided 4) At 2 minutes, take a picture of the test chart with your phone's camera, using a dedicated application downloaded beforehand 5) Send the picture, and the result appears almost instantly.

Advantages: A relatively inexpensive, rapid procedure that enables the effects of agricultural practices to be observed and corrective measures to be taken if necessary.

Limitations: Simplified detection can only indicate large masses, and can under no circumstances provide information on fungal species or bacterial strains.

16.7 Weeds as bioindicators of soil fertility

Operating principle: Bioindicator plants are plants that cope better with certain imbalances, excesses or deficiencies than crops. Their establishment and prosperity reflect specific physico-chemical soil conditions (Chikishev, A. G. 1965 and Sampson, A. W. 1939). Their better adaptation gives them a competitive edge over crops. Reducing these excesses or imbalances will put the crop back in the best possible position to assert itself, and recreate the conditions for selective herbicides to regain their effectiveness.

How it works & conditions for validating weeds as soil nutritional indicators: Bioindicator plants are a warning signal that should alert the agro-ecologist. We can use them as a decision-making tool for mineral rebalancing. However, the presence of just a few plants of a species described in the table below is not enough to identify, without fail, the mineral elements that may be deficient or in excess. For this, there must be :

1. sufficiently dense populations, covering more than 10% of the surface area
2. “aggressive” weeds that are highly competitive with the crop
3. several species with the same interpretation keys present to enable a sufficiently reliable conclusion to be drawn

Advantages: Free tool. As a general rule, the addition to the soil of the elements determined by the indications in the table below (backed up by soil analysis if necessary), makes weed competition less pressing.

Limitations: The analysis of bioindicator weeds needs to be complemented by a study of soil functioning and soil analysis. It should be noted that bioindicator plants have the particularity of exploring a much greater depth of soil than the horizon from which soil samples are taken. This point alone may explain some apparent contradictions.

How it works: This table primarily serves as a list of the physicochemical characteristics to be checked in each situation where a weed is present, knowing that all or only part of the listed excesses/deficiencies will be present.

The (-) indicates low bioavailability in the soil. The (+) indicates an abundance, or even a possible excess, of elements in bioavailable form. The (+) for aluminum (Al+) signifies that the element's abundance exerts toxicity on cultivated plants, thus creating an ecological niche favorable to weeds that can tolerate it. Empty cells indicate that no link has been identified between the presence of the plant and the associated mineral. We have marked with a cross (X) the situations where the Ca/Mg ratio is unfavorable, and also with a cross (X) the contexts where anaerobic processes (suffocation) are likely predominant. These indications are a compilation of numerous sources (*Hill, Stuart & Ramsay, Jennifer. (1977*). *Weeds as Indicators Of Soil Conditions*. MacDonald J.. 38 - "*When Weeds Talk*" by Jay L. McCaman - "*Weed Control without Poisons*" by C. Walters/Acres USA - "*Weeds and What They Tell*" by Ehrenfried E. Pfeiffer) - Holzner and Namata, 1982 - Chauvel & Al 2001 - and our own experience.

Table : Correspondence Between Weeds and Mineral Deficiencies

Common name	Latin name	Ca/Mg	Ca	P	K	Mg	Mn	Fe	S	Cu	Zn	B	Cl	Se	Anae Bac	Al+	NO3
American Pokeweed	Phytolacca americana	X	-	-	+		+	+			+	+		+	X		
Annual Bluegrass	Poa annua	X			+	+	+	+	+		+				X	+	
Barnyard Grass	Echninochloa crus-galli	X	-	-	+		+		+	+	+				X		+
Black Nightshade	Solanum nigrum	X	-	-	+		+	+	+			+	+	+	X		+
Black-bindweed	Polygonum convulvus	X	-	+	+			+	+	+				+	X	+	
Bracken Fern	Pteridinum aquilinum		-		+						+					+	-
Broadleaf Plantain	Plantago major		+	+	+		+	+			+	+	+	+	X		
Broad-leaved Dock	Rumex obtusifolius	X	-	-	+		+	+		+	+	+	+	+	X	+	+
Brown Knapweed	Centaurea jacea	X	-	-	-	+								+	X	+	
Burdock	Arctium		-					+									
Canada Thistle	Cirsium arvenses		-	+	+		-	+		-					X		
Catchweed Bedstraw	Galium aparine	X			+												+
Charlock/Wild Mustard	Sinapis arvensis	X	+	+			-	-							X		+
Chickweed	Stellaria media	X	-		+	+		+	+	+	+	+	+	+		+	+
Common Cocklebur	Xanthium strumarium	X	-	-	+	+	+			+	-		+	+	X		+
Common Fumitory	Fumaria officinalis	X	+		+								+				+
Common Groundsel	Senecio vulgaris	X		-				+						+		+	+
Common Hemp-nettle	Galeopsis tetrahit	X	-	-													
Common Purslane	Portulaca olearacea	X	-	-	+	+		+	+	+					X		
Common Ragweed	Ambrosia artemisiifolia	X	-	+	-	+			+	-							-
Common Sorrel	Rumex acetosa	X	-	-				+								+	

Common name	Latin name	Ca/Mg	Ca	P	K	Mg	Mn	Fe	S	Cu	Zn	B	Cl	Se	Ana e Bac	Al+	NO3
Common Tansy	Tanacetum vulgare	X	-	-	+		+			+							
Common Thistle	Cirsium vulgare	X	-	+			-	+			+	+		+		+	
Corn Buttercup	Ranunculus arvensis	X	-	-	+		+									+	
Corn Chamomile	Anthemis arvensis	X		-	+						+	-			X	+	-
Corn Spurrey	Spergula arvensis	X	-	+	-		+	+	+	+	+	+	+	+			
Cornflower	Centaurea cyanus	X	-	-	+	+	+	+	+	+	+	+	+	+			+
Couch Grass	Elymus repens		-												X	+	
Crabgrass	Digitaria sanguinalis	X	-	-	+			+	+	+	+	+	+	+	X		+
Creeping Buttercup	Ranunculus repens	X	-		+										X	+	
Creeping Soft Grass	Holcus mollis		-		-										X		-
Curled Dock	Rumex crispus	X	-	-	+	+	+	+	+						X	+	
Cut-Leaved Crane's-Bill	Geranium dissectum		-	-	+			+								+	+
Dandelion	Taraxacum officinale	X	-		+								+		X		+
Dodder	Cuscuta	X		-	+												
Field Bindweed	Convulvus arvensis	X	-	-	+	+									X		+
Field Gromwell	Lithospermu m arvenses	X	+	-	+						+						
Field Horsetail	Equisetum arvense		-	-									+				
Field Mint	Mentha arvensis		-	-			+	+							X		
Field Pansy	Viola arvensis		-				-								X		-
Field Pennycress	Thlaspi arvense	X	-	+	+	+	+		+				+	+	X		
Field Pepperwort	Lepidium campestre	X	+	-	+	+	+		+		+	+	+	+			
Field Sedge	Carex caryophyllea	X	-		-	+		+			+	+		+		+	
Foxtail	Alopecurus	X													X		+
Foxtail Barley	Hordeum jubatum	X	-		-										X		+
Great Mullein	Verbascum thapsus	X	-	-	-				+		+	+		+		+	
Green Foxtail	Setaria viridis	X			+	+		+	+		+	+	+	+	X		+

Common name	Latin name	Ca/Mg	Ca	P	K	Mg	Mn	Fe	S	Cu	Zn	B	Cl	Se	Anae Bac	Al+	NO3
Ground Ivy	Glechoma hederacea		-					+	+	+	+	+	+	+			
Hairy Rockcress	Arabis hirsuta	X	+			-	+	+			+			+	X	+	-
Hedge Bindweed	Calystegia sepium	X	-	-	+	+		+	+						X	+	+
Hedge Mustard	Sisymbrium officinale	X	+	-	+	-	-		+	+	+	+		+	X		+
Italian Ryegrass	Lolium multiflorum	X	-		+			-							X		+
Jerusalem Oak Goosefoot	Chenopodium botrys	X	-	+			-	-		-		-			X		
Jimsonweed	Datura stramonium	X	-	+	+	+		+	+		+	+		+	X		
Johnson Grass	Sorghum Halepense	X	-		+	+	+	+	+	+	+	+	+	+	X	+	
Knotweed	Polygonum aviculare	X	-	+	+	+								+	X	+	+
Lamb's Quarters	Chenopodium album			+	+				+								+
Leafy Spurge	Euphorbia esula	X	-	-	+	+		+	+			+		+	X		
Meadow Buttercup	Ranunculus acris	X	-	-	+	+	+	+	+	+		+		+	X	+	+
Mexican Tea	Chenopodium ambrosoides	X	-	+	+		+				+	+	+		X		
Mugwort	Artemisia vulgaris	X	+												X		-
Narrowleaf Plantain	Plantago lanceolata	X	-	+	+	+			+	+		+	+	+		+	
Oak-Leaved Goosefoot	Chenopodium glaucum		-	+					-	-							
Oxeye daisy	Leucanthemum vulgare	X	-	-	+	+											
Peavine	Lathyrus	X	+	-	+												
Perennial Ryegrass	Lolium perenne	X				-						-					+
Perennial Sowthistle	Sonchus arvensis	X	-		-									+	X		+
Prostrate Pigweed	Amaranthus graecizans	X	-	+									+				+
Purple Deadnettle	Lamium purpureum	X	-	-	+												+
Redroot Pigweed	Amaranthus retroflexus	X	-	+	+	+				+	+	+	+	+			+
Roof Brome	Bromus tectorum	X	-	-	+								+	+	X		-
Rye Brome	Bromus secalinus		+	-								-					

Common name	Latin name	Ca/Mg	Ca	P	K	Mg	Mn	Fe	S	Cu	Zn	B	Cl	Se	Ana e Bac	Al+	NO3
Scarlet Pimpernel	Anagalis arvensis		-	-													+
Scentless Mayweed	Matricaria recutita	X	+		+												+
Sheep's Sorrel	Rumex acetosella	X	-	-										-		+	
Shepherd's Purse	Capsella bursapastoris	X	+	+	-		+	+	+	+	+	+	+	+	X	+	+
Silverweed	Potentilla anserina	X	-	+					+			+	+	+	X		
Slender Thistle	Carduus tenuiflorus	X	+	+		+										+	
Small Nettle	Urtica urens	X	-	-	+	+				+							+
Smooth Pigweed	Amaranthus hybridus	X	-	+	+	+	+	+	+	+		+	+				+
Soft Brome	Bromus hordaceus		-	-			-	+				-			X	+	+
Spiny Sowthistle	Sonchus asper	X	-	+	+				+			+			X		+
Sterile Brome	Bromus sterilis	X		-	-										X		
Sterile Brome	Bromus sterilis	X	-			-									X	+	
Stinging Nettle	Urtica dioica	X		+	+	+	+	+	+	+	+	+	+	+	X	+	+
Upright Brome	Bromus erectus	X	+														-
Velvetleaf	Abutilon theophrasti	X	-	-	+	+	+	+	+	+		+	+	+	X	+	
Viper's Bugloss	Echium vulgare	X	-		+		+	+	+		+		+			+	
Wall Speedwell	Veronica arvensis	X			+								+				+
White Mustard	Brassica alba	X	+	+	+		-		+	+		+	+	+			+
Wild Carrot	Daucus carota		-	-	+			+				+	+	+	X		
Wild Garlic	Allium vineale	X	+		+			+				-	+		X		-
Wild Oat	Avena fatua	X	+	-	+		+			+	+	+		+	x		+
Wild Parsnip	Pastinaca sativa	X	-	-	+	+			+						X		
Wild Teasel	Dipsacus fullonum	X	-	-	+	+		+	+		+	+		+		+	
Yarrow	Achillea millifolium		-	-	-		+		+							+	-
Yellow Foxtail	Setaria glauca	X	-		+	+									X	+	+

Major factor

The calcium/magnesium ratio to curb weeds

When it comes to weeds, we can see how important and predominant the calcium/magnesium ratio is. This ratio must first be checked and then resolved at the soil level. In addition to excessively high or low mineral values, other major factors influence weed establishment: soil structural condition, soil waterlogging, organic matter that is either prematurely mineralized or excessive but lacks proper evolution. There are well-documented works on these points that can be usefully consulted, such as Gérard Ducerf's "Encyclopédie des plantes bio-indicatrices (Encyclopedia of Bioindicator Plants)" published by Promonature.

17. IMPROVEMENT STRATEGIES FOR BIOLOGICAL, PHYSICAL, CHEMICAL AND WATER FERTILITY

17.1 Biological fertility: restoring microbiological balance

As we have seen, it is biological activity that is responsible for the creation of soils, in all their mineral and organic components. The balance of the action of the many families of microflora and microfauna, bacteria, and fungi is crucial in the evolved soils on which we work. Humus serves both as shelter and nourishment. Soil restoration involves the production of precursor materials for humus and the mobilization of the most suitably adapted microbiological agents.

17.1.1 Methodology: how to establish a balanced diet?

The management of organic inputs to the soil deserves to be considered in the same way as a complete diet for ruminants. Imbalanced material returns should never be delivered to the soil in a staggered or asynchronous manner. One does not give protein concentrates on Monday, coarse forage on Tuesday, and minerals on another day. As with a dairy cow's ration, everything must go through the "mixer wagon"; otherwise, a ration that may be balanced on an annual average will generate successive imbalances. Even if they are only a few weeks apart, this will never produce a beneficial effect. In a large-scale farming system, the return of crop residues rich in cellulose should accompany cover crops that are too rich in nitrogen and sugars. To achieve this, cereals should be harvested as high as possible, collecting only the ears, and green manure should be sown in the stubble. To continue the ruminant metaphor, here the food storage shed will be the field with the standing straw, and the mixer wagon will be the cover crop or the FACA roller. Only mulch everything through soil work during the destruction of the cover crop. There will then be a nutritional synergy for the soil microflora, and the soil fungi will benefit. Moreover, the standing straw will have had plenty of time to wash away its anti-growth substances and will neither hinder the cover crop nor the main crop.

The mechanical treatment of straw should not result in too fine a grinding. Indeed, this could create a mat pressed to the ground, causing asphyxiation of the surface horizon in wet weather and serving as a refuge for slugs. It is better to superficially incorporate long straw pieces for rapid decomposition than to leave a mat of short pieces on the surface (Fukuoka method).

17.1.2 Expected benefits

- Limitation of surface acidity by fixing bases (notably by pseudomycelia) thanks to the fungi regaining their place. Neutralization of excess minerals (Fe, Mn...)
- Improvement of fertility through better bioavailability of elements.
- No risk of "nitrogen starvation".
- Neutralization of the allelopathy of cereal straw.
- Improvement of surface structure.
- A more resilient soil environment less sensitive to climatic hazards.
- Improvement of crop health.

17.1.3 Seeding or targeting beneficial microbial flora

■ Context

Agricultural practices have tended to limit the action of certain representatives of soil biological activity. Particularly, the aerobic rhizospheric microflora becomes vulnerable at crop maturity. Indeed, in an unbalanced environment, indigenous flora, which are adapted to the environment, will eliminate the strains that are too rhizodependent.

■ Methodology

Use of PGPR strains (*Plant Growth-Promoting Rhizobacteria*):

- Reintroduction of organic fertilizers (see above).
- Implementation of targeted activators containing trace elements that activate their enzymes (see Chapter 19 on fertilization).

■ Expected Benefits (depending on the types of microorganisms inoculated into the soil)

- Restoration of soil resistances to diseases (e.g., flea beetles on flax, etc.).
- Increase in nutritional density, sugars, and minerals.
- Degradation of retrograded forms of minerals and alteration of the primarily mineral C horizon.
- Release of fertile elements from the soil in a bioavailable form.
- The "microbial bridge" established by this practice produces

productivity gains even in chemically "rich" soils. Indeed, during certain periods highly favorable to plant growth (heat, humidity, day length), microorganisms (*bacillus mucilaginosus* being particularly effective in this regard) accelerate the flow of minerals from the soil to the roots. The production of dry matter is no longer limited.

17.1.4 Optimizing diazotrophs

The fixation of nitrogen from the air by bacteria (symbiotic or non-symbiotic) is a functionality of agroecosystems that can be interesting to activate. These bacteria may have suffered from poor soil conditions. Re-inoculating the soils can be beneficial. Generally, rhizobia associated with legumes remain in the soils after the strains adapted to the cultivated species have been introduced. Free nitrogen-fixing bacteria are very interesting because they are not specific to a species (for example, *Azotobacter chroococcum* or *vinelandii* can adapt to the rhizospheres of fruit trees, field crops, and oilseeds) and can provide up to 30% of the nitrogen needs of plants. They require annual re-inoculation. However, it should be noted that these bacteria have specific needs that must be met for successful use:

- **Maximum energy efficiency.** They manage to fix nitrogen at ambient temperature, a process that the Haber-Bosch method (which produces nitrogen fertilizers from fossil hydrocarbons) achieves under extreme temperature and pressure conditions and is therefore very energy-intensive. To accomplish this (N_2- NH_2-NH_3), energy transport within the bacteria is crucial and involves phosphorylated molecules like ADP- and ATP(Adenosine diphosphate and triphosphate). The phosphorus level in the soil is therefore essential (Kashif, 2012). (See chapter on mineral fertilization)
- **Optimization** with nitrogenase catalysts: molybdenum and cobalt
- **Sulfur**: All nitrogen-fixing bacteria need sulfur to participate in the production of certain amino acids (lysine, methionine, etc.) and to build their own tissues. Nitrogen is an unlimited source for diazotrophs: there is a column of air containing 70,000 tons of dinitrogen (N2) above one hectare of land. However, sulfur, phosphorus, and necessary trace elements are limited by the soil's content. Therefore, sulfur levels must also be sufficient. This is a particularly important point of attention in sulfur-exporting systems (e.g., legumes).

Conclusion: For the nitrogen-fixing function, whether by symbionts (nodules) or by free bacteria, to be active in the soil, minimum levels of phosphorus, molybdenum, cobalt, and sulfur in bioavailable forms are necessary. These levels should be measured and corrected if necessary.

17.1.5 Promoting mycorrhizae

Mycorrhizae are objective allies of 95% of cultivated plants. As such, we must create favorable conditions that allow them to thrive.

Mycorrhizae: A nutrient accelerator for plants

> Fungi cannot perform photosynthesis. However, through their mycelium, they have an amplified access to nutrients. Thus, they form associations with plants. By providing nutrients to the plants, they receive sugars in return. The plant helps the fungus grow, and the fungus helps the plant grow. The faster the plant grows, the more the fungus expands and nourishes the plant. A virtuous circle is established. As a result, mycorrhized plants benefit from an absorption potential up to ten times greater. Another beneficial effect of mycorrhizae for plants is their protective effect on roots against nematodes and other pathogens (Beck Malcolm, 2005).

The necessary conditions for their establishment are largely described in Chapter 19 on fertilization: Ca/Mg ratio, phosphorus level, among others. They are much more sensitive to excessive mineral richness than to deficiencies. This is why fertilization programs must be conducted with the utmost caution. Furthermore, fungicides are generally harmful to them, so it is advisable to reduce their use in crop rotations. Finally, a few major crops do not host mycorrhizae, such as rapeseed, cabbage, and beets, and their frequent presence in a rotation can greatly influence the mycorrhization of subsequent crops. Studies have shown that corn following rapeseed, fallow, or beets had less mycorrhized roots compared to corn following sunflower, soybeans, potatoes, corn, or wheat, with rates ranging from 2% (rapeseed) to 100% (Arihara & Karasawa, 2000). Yield variations correlated with the mycorrhization rate. This effect can be further accentuated if crucifers are frequently used as cover crops (mustard, forage rapeseed, turnips, etc.).

It thus appears that mycorrhization is much more dependent on fertilization practices, crop rotations, and the use of fungicides than on direct inoculation. Mycorrhizal inoculation can be considered once at the beginning of a revitalization program, but if the main guidelines described here are followed, the soil will maintain a lasting mycorrhizal potential.

17.1.6 Reducing fungicide programs

The primary agents capable of 'producing' stable carbon are fungi, mycorrhizae, ascomycetes, and basidiomycetes. Improving this soil functionality, which increases organic matter content, involves reducing fungicide programs. As their name suggests, fungicides have a marked effect not only on pathogens but also on beneficial soil fungi. Our observations have led us to conclude that the last fungicide applied is much more persistent and impactful. This persistence is particularly strong when the application is followed by a dry spell, preventing the dilution and degradation of pesticides. This results in very high residual levels in the straw. The best way to avoid or significantly reduce the doses of the last fungicide, without risk, is through perfect plant nutrition via the mineral and microbiological balance of the soil. We can call this nutriprotection. This is the first line of defense.

Plant tea (or compost tea) should be used knowingly and with caution

> The use of anaerobically-produced, foul-smelling plant-based teas can certainly be effective in controlling certain diseases and pests. However, given their concentration in reduced elements (ammonia, sulfides, aldehydes, etc.), they are likely to encourage the activity of the least desirable families of microorganisms. On the whole, anything that smells bad is detrimental to the biological activity of the soil or phyllosphere.

17.1.7 Cover crops

Many consultants and trainers are true specialists on the subject and can only make a useful contribution to the appropriate and adapted use of cover crops for each situation.

The use of cover crops is of major utility for the biological activity of soils, maintaining porosity, and resistance to erosion (see section 17.2 below). The results are more mixed regarding sustainable carbon storage if the conditions described repeatedly in the book on stages, methods of destruction, and incorporation are not followed.

17.2 Levers for improving soil physics

Mechanical operations will never replicate what living organisms are capable of doing. However, mechanical action can help resolve certain dysfunctions or maintain particularly fragile soils. In this section, we will describe a few situations where resorting to heavy operations, sometimes considered outdated, can be beneficial.

17.2.1 Agronomic tillage

■ Context

In the specific cases of sandy-loam and loamy-sand soils, we recommend considering a return to or maintenance of plowing. Indeed, we have seen how structurally fragile these soils are: no earthworms and no swelling clays. They also have a propensity to lose their surface cations. The justified promotion of direct seeding and simplified tillage techniques makes us too quickly forget some advantages of agronomic plowing. The main disadvantages attributed to plowing are those related to deep plowing. In certain situations, it was not the plowing itself that needed to be reconsidered, but only the depth.

■ Methodology

In order not to mix soil horizons, these plowings should be shallow, with plowshares (with a working width of 10-14 inches (34 cm) maximum) without skimmers to avoid burying crop residues at the bottom of the furrow. These shallow plowings do not create a plow pan because the work is done in a zone still strongly influenced by climatic conditions. The plowing should not exceed a depth of 12 to 18 cm. It is worth noting that our ancestors plowed very superficially, at 10-15 cm maximum. The plowshares have gradually widened only to accommodate the increasingly wide tires of more powerful tractors. This is how deep plowing was born.

For agronomic plowing, the plow used should have plowshares of a maximum of 12 inches to allow proper soil turnover without the need for deep work. The plow should not have skimmers to avoid embedding fresh organic matter at the bottom. Finally, the plow should have a large number of shares to enable off-furrow work. A shallow off-furrow plowing does not cause a plow pan: the weight transfer from the moldboard to the heel is much lower, there are no tractor wheels at the bottom of the furrow, and any compactions are at a depth where climatic impacts still have significant beneficial effects: freeze-thaw cycles, dry-moist alternation, and plant root density.

These agronomic plowings require adopting agronomic plows, sometimes still referred to as stubble plows. Manufacturing has been revived by various manufacturers (Bonnel, Bugnot, Charlier, Goizin, Kverneland, etc.). The benefit of shallow plowing includes savings in fuel and reduced wear. Other advantages include mineral recycling (including calcium, prone to leaching losses), combating surface acidity, and continued effective use of root herbicides, thereby facilitating weed control. It should not be forgotten that weed control was the primary motivation for our ancestors to plow.

A plow without skimmers?

> Coulter discs should be avoided. Indeed, with today's wide plowshares, crop residues are thus transported to depths where the lack of oxygen leads to anaerobic degradation and the establishment of anoxic conditions. However, skimmers may find their place on agronomic plows and in organic practices, for example when no herbicide program is planned. Their action can then be beneficial in bringing straw, and especially weed seeds, to a reasonable depth and thereby limiting their germination.

17.2.2 Subsoiling - cracking - decompacting

■ Context

The formation of plow pans or no-till pans is a phenomenon that generally does not resolve itself. Without intervention, they often tend to worsen from year to year. In the presence of highly compacted pans, experience shows that mechanical intervention is necessary to restore soil function within a reasonable timeframe and cost for the farmer. Soil self-repair, even with the help of cover crops, does not always occur. Based on our diagnostics, the observation of cover crop root system architecture leads us to conclude that they are more indicators of compaction rather than decompactors. Therefore, the use of a subsoiler should be considered in these situations.

■ Methodology

Subsoiling must be carried out by precisely adapting the working depth so that the tines of the tool can work just below the compacted layer. Above all, we must not degrade the deep horizons which generally have an excellent structure. A chronology is essential in decision-making: diagnose the need to use it by conducting a soil profile and then precisely identify the necessary depth. Make a round trip with the subsoiler then ensure that the work is as desired (in particular the working depth). Loosening the soil at a deeper level than necessary involves the deterioration of an original stable porosity and expending more energy than necessary. Loosening the soil at a shallower level than desired will involve transporting water into the lowest parts of the field and thus creating wet areas that did not exist before.

■ Expected benefits

- Deep water recirculation, reduction of waterlogging, and elimination of perched water tables.

- General restoration of soil verticality: rapid root penetration.
- Restoration of mineral nutrition and elimination of "shadow zones" (areas not explored by roots that bypass compacted zones).

■ **Possible negative effects and points of caution**

- The risk of systematic and overly frequent use of a subsoiler is soil compaction. It is common practice among public works companies to use a subsoiler as a preliminary step to better compact soil. Indeed, a soil whose internal architecture has been pulverized is much easier to compact. It's akin to the effect of a missile on a building: it becomes much shorter, less porous, and leaves less space for its occupants after the explosion.
- In sandy or loamy soils, there is a risk of "knocking down" fine particles, including clays, to the bottom of the working depth when working under very dry conditions.
- "Biological" cracking: while many cover crops can maintain adequate porosity, few can truly restore highly degraded soil structure.
- **Subsoiling: as little as possible, but as often as necessary.**

The loosening power of tall fescue

> Highly compacted plow pans cannot be cracked by commonly cultivated non-woody plants. However, there is an exception in temperate climates: tall fescue. This crop, however, is not easy to establish due to its small seed that germinates poorly. Its market value is low, and its effectiveness requires occupying the soil for a minimum of eight to twelve months. Its implementation may be interesting on a case-by-case basis, especially when subsoiling is impossible (such as in orchards, for example). In certain situations, alfalfa can play a comparable role.

17.3 A more global approach to chemical fertility

The cultivated ecosystem is rarely able to completely recycle soil cations, even if the actions mentioned earlier have been implemented. It is therefore common to consider corrective actions through external inputs.

17.3.1 Expanded decision matrix for mineral corrections

■ **Methodology**

Analysis of the soil profile provides a highly valuable source of information for determining soil amendments. pH measurements at various points in the profile inform us about the extent of the soil volume impacted by acidification and decarbonation. Amendments are calculated based on these affected volumes. Amendments can be reduced when deep horizons show high calcium levels and it is observed that plants have access to these deep layers. Conversely, if there is a decreasing gradient, calcium and/or magnesium inputs should be considered not only to correct the surface layers but also the entire soil column below.

Note: In complex situations, compromises have been adopted based on soil profile observations. For instance, surface sandy soils overlaying heavy clays at 60-80 centimeters depth might necessitate using Ca/Mg ratios of 60-20 (as discussed in section 19.2) to improve soil cohesion. However, in the long term, this could make the clay horizon even more cohesive and sticky (resulting in slower water infiltration, formation of perched water tables, and likely reduced root depth). In such cases, a compromise is adopted by applying magnesium in moderate quantities, well below the model presented in section E.

■ **Expected benefits**

- Improvement of soil structures in both surface and deep horizons.
- Fixation of free clays: The addition of cations allows "free" clays to combine with humus, flocculate, and thereby become less prone to leaching.

17.4 Water fertility: how to optimize it?

17.4.1 Mechanical operations

Drainage and irrigation: Cracking can partially or completely alleviate the effects of a "perched water table" and extend the intervention windows in the fields. This operation also allows for increased water storage and provides unrestricted root access.

17.4.2 Ca/Mg balances

Good fertilization practices (section 19.2) will restore the proper air-water proportions in the soil. This promotes moisture retention, root density, and thus optimal yield for crops. The potential for holding water (field capacity) can be significantly enhanced by paying attention to this ratio.

17.4.3 By reconstituting humus levels

This results in improved water storage capacity (up to 20 times its own volume) and, importantly, during mineralization, ensures a balanced release of minerals into the soil solution that closely matches the plant's needs. This promotes efficiency in the use of available water.

17.4.4 Through biological fertility

The implementation of the previously reviewed measures: reactivation of biological activity is essential for optimizing soil capacity for storage, but also particularly through mycorrhizae, for enabling crops to access stored water.

E

Restoring soil fertility

18 • SOIL ANALYSIS: MAKING THE MOST OF IT

Today, we observe in the field that many farmers claim not to "believe" in soil analyses anymore. Although soil analysis does not replace the detailed knowledge of one's soil gained through cultural profiling, it complements it significantly. A well-interpreted soil analysis can be a crucial and irreplaceable decision-making tool. We believe that the current distrust towards soil analyses stems partly from a sometimes misunderstood framework. Indeed, laboratory methodologies incorporate various biases that are seldom explained or accounted for. These biases can originate from sampling practices themselves and the analysis methods employed. Furthermore, some criteria are not directly measured but rather calculated.

In this chapter, we aim to provide you, criterion by criterion, with key insights to better interpret soil analysis results.

18.1 Objectives and methods for relevant analyses

18.1.1 Method: taking into account and limiting biases

Common soil analyses provide numerous numerical indications. Some are direct measurements, while others are calculations derived from formulas applied to the measured elements. We will distinguish between them in the following descriptions. It is crucial to consider that soil analysis is and should remain a decision support tool, often providing indicative values. Even though the reported values are precise to parts per million (ppm), parts per gram (gr), 0.1 pH unit, or 0.1 milliequivalents (meq), the margins of error often exceed these units by a significant margin. These errors are unavoidable and do not undermine the usefulness of laboratory measurements.

Variability stems from several factors, including the reliability of sampling, the timing of sampling, recent applications of mineral and organic amendments, laboratory processes, and instrument calibration.

18.1.2 Good sampling practices

No laboratory procedure can correct for a flawed sampling methodology. For each sample, ideally, about fifteen soil cores should be taken, preferably based on a systematic grid or using fixed landscape elements, or better yet, by geolocation. This method allows for more precise monitoring of fertility changes. Subsequently, it is essential to thoroughly mix the samples to obtain a laboratory sample that is as representative as possible.

The sampling depth must be consistently uniform. For annual crops under conventional tillage, a depth of 20 cm on packed soil is ideal. To establish a long-term reference on a farm without introducing the variability of random depths, it is crucial to sample each year at exactly the same depth. This consistency is vital unless transitioning to no-till practices, which are increasingly common and described below. It is imperative to equip sampling tools (such as augers or gouges) with a depth stop (adjustable if necessary) to ensure the chosen depth (5 cm, 20 cm, 60 cm) is maintained.

18.1.3 Specific cases: minimum tillage and direct seeding

Given that minimum tillage (MT) or no-till practices no longer mix and homogenize soil layers as traditional plowing did, specific precautions must be taken when conducting soil analyses. The concentration of organic matter at the surface, which enhances structural stability, can lead to significant errors in interpreting soil analysis values. The two indicators most sensitive to this phenomenon are organic matter content (which is concentrated at the surface) and phosphorus (which is relatively immobile in the soil). In these situations, where there has been a recent or ongoing transition from plowing to a non-inversion tillage system, **two types of sampling approaches are recommended:**

1. **A sampling depth that matches the plowing depth previously practiced** (often 30, 35, 40 centimeters, or more) is necessary to establish meaningful trends in both organic and chemical soil fertility. This method allows farmers to monitor their fields consistently and enables laboratories to maintain comparative values with historical data. Failing to sample at this depth typically leads to misleading self-satisfaction, as it may not promptly detect overall fertility depletion or may overestimate certain parameters like soil organic matter. Even if the sampling depth during conventional tillage was not directly related to the plowing depth, the values obtained were generally representative of the entire cultivated volume due to the minimal homogenization achieved through annual operations. This is precisely why we still need to base our assessments on this consistent depth today.

2. **A second sampling should be done at a depth between 0 and 5 centimeters.** This zone is particularly susceptible to significant loss of surface cations and accelerated acidification due to the absence of plowing, which no longer brings these cations back to the surface (calcium, magnesium, potassium, sodium). In no-till situations, it is common to find water pH levels as low as 5.5 in this zone, even in soils supposedly saturated with calcium, where a soil analysis taken from a 30-centimeter depth might show a pH of 7.5 or 8. This becomes problematic because it's the zone where seeds are placed and the only environment where young seedlings must establish their emerging root system (spermosphere).

Another potentially concerning effect is that this is where most of the organic matter returns are deposited (aside from that contained within the root system). Without plowing, the evolution of this organic matter is confined to this undersaturated environment. Yet, the quality of humus derived from this organic matter transformation is highly dependent on pH and the presence of cations. Its half-life and cation exchange capacity (CEC) are influenced accordingly, affecting whether a high CEC mull-type surface horizon is created or a more moderate CEC moder/mor horizon with a shorter lifespan is formed.

Figure 18.1 clearly illustrates how non-tillage practices significantly alter the distribution of organic matter within the soil column, thus justifying a distinct sampling methodology.

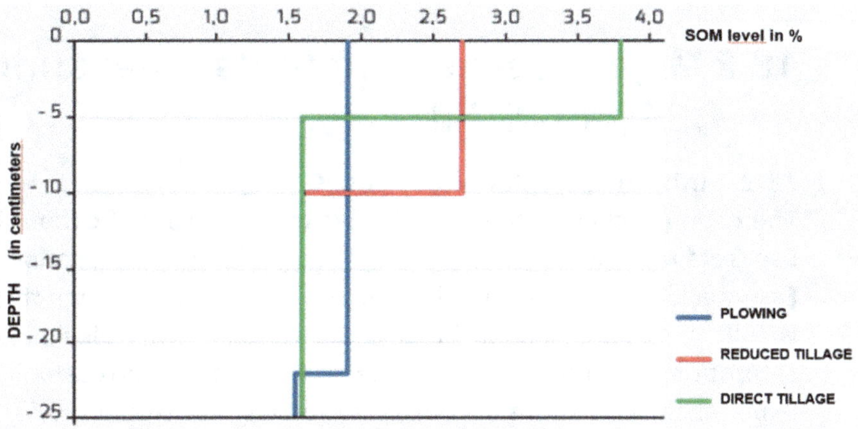

Figure 18.1 – Evolution of total carbon content in the soil profile in an ITCF plot in Boigneville with 3 treatments: conventional tillage (solid line), conservation tillage systems (CTS), and no-till (dotted line) over 28 years (Labreuche et al., 2001).

18.1.4 Deep sampling according to soil conditions

Another level of sampling between 60 centimeters and 2 meters deep can be considered to reach a different deep layer identified via soil profile. Depending on soil conditions, this can be highly beneficial at least once in a producer's career, especially **for perennial crops such as vineyards, fruit trees, hops, and olive trees,** which have deep root systems capable of absorbing nutrients from deep layers. Deep soil analyses are also recommended in large-scale farming where **geological disturbances exist**. Agricultural soils may have surface horizons that are not necessarily derived from underlying layers—for example, alluvial plains, glacial moraines transported hundreds of kilometers from their origin, aeolian loams in coastal regions of northern Europe, or marine deposits in Dutch polders, among others. Significant differences can exist between the worked layer commonly analyzed and deeper layers explored by roots. This knowledge can greatly influence fertilization recommendations and lead to substantial input savings. For instance, it may be relevant to apply nutrients only locally at seeding to ensure they are available during early stages of growth if the subsurface is rich in the same element. In such cases, additional practices like retaining crop residues on-site and using deep-rooting cover crops can help transport elements from deep soil layers (such as calcium, magnesium, phosphorus, and trace elements) that are deficient near the surface. While these supplementary measures may not fully restore optimal chemical fertility, they contribute to the profitability of farming systems by leveraging all available resources of the land.

18.2 Major assessment tools : measuring CEC, pH, SOM

18.2.1 Introductory definition of CEC

The cation exchange capacity (CEC) is the total sum of cations that a soil can retain reversibly on its negatively charged adsorption complex. Essentially, CEC represents the soil's cation reservoir. When this reservoir is full, the CEC is said to be saturated (common in alkaline soils). The saturation percentage of CEC indicates the level of this reservoir. From our experience, we believe that a saturation level of around 90% is generally optimal. CEC is primarily influenced by clay minerals and humus. Due to the laboratory determination methods for clay and silt fractions, some "silt-sized particles" (which are actually coarse mineral clays) also contribute to CEC.

Cations are temporarily fixed on the CEC and are not structurally part of clay minerals or organic matter. Therefore, they can easily be released back into the soil solution to participate in plant nutrition. This property helps regulate the soil solution composition. In temperate climates, towards the end of the growing season when microbial activity decreases in autumn, minerals released, including ammonium forms from microbial proteins (such as NH_2^+ and NH_4^+), are stored rather than quickly oxidized to nitrates and leached during winter. Therefore, high NH_4^+-N values in winter soil analyses often indicate intense biological activity during the warm season, which is a positive sign. CEC values are typically expressed in milliequivalents per 100 grams of soil (sometimes now in meq/kg). Extreme values can range from 2 meq/100 g to 50 meq/100 g.

The distinction between meq (milliequivalents) and mg (milligrams) is important because these units have very different meanings for each of the cations involved ; mg (milligrams) is a unit of weight, while meq (milliequivalents) represents a unit of electrical charge. These two units are not directly comparable because they depend on the molar mass and valence (number of electrical bonds possible) of each element. For instance, 120 kg of magnesium mobilizes the same amount of electrical charge on the CEC (Cation Exchange Capacity) as 390 kg of potassium (see Table 17.1). In other words, potassium requires much more weight compared to magnesium or calcium to saturate the same cation exchange capacity, as illustrated in the table below:

Table 18.1 – Table expressing, by cation, the mass share in ppm, mg/100 g of soil, and kg/1,000 t of soil required to occupy 1 meq of CEC.

Cation: meq/100 g unit of electrical charge	Mass share (in ppm)	mg/100 g of soil	kg/1,000 t of soil required to occupy 1 meq of CEC
Calcium 1 meq	200 ppm	20 mg	200 kg
Magnesium 1 meq	120 ppm	12 mg	120 kg
Potassium 1 meq	390 ppm	39 mg	390 kg
Sodium 1 meq	230 ppm	23 mg	230 kg
Hydrogen H+ 1 meq	10 ppm	1 mg	10 kg
Aluminum 1 meq	90 ppm	9 mg	90 kg

■ **Conversion to pure elements**

Soil analyses provide the quantities of minerals found in their oxidized form. To understand a soil analysis, it can also be useful to convert elements expressed as oxides into pure elements. Here is a conversion table.

Table 18.2 – Conversion table of oxide mass to pure cation mass

Oxides	Conversion rate	Pure elements
CaO	× 0.715	Ca
MgO	× 0.603	Mg
K_2O	× 0.830	K
Na_2O	× 0.742	Na
P_2O_5	× 0.44	P

Anion exchange capacity

It should be noted that in some soils, there is a significant AEC (anion exchange capacity). This is the case for andosols of volcanic origin. These soils have the ability to retain anions ($BO^-{}_3$, $NO^-{}_3$, $SO_4{}^{---}$).

The main cations present in the clay-humus complex, constituting the Effective Sum of Cations:

$$T\ eff. = Ca^{++} + Mg^{++} + K^+ + Na^+ + Al^{3+} + H^+$$

The cation $NH_4{}^+$ is also present and adsorbed like the other cations, but it is so minor that it is generally not taken into account.

■ **Exchange Acidity Rate**

The cations Al^{3+} and H^+ are mainly present in unsaturated, thus acidic soils. The sum of these cations Al^{+++} and H^+ expressed as a percentage of the total effective CEC is called the EAR (Exchange Acidity Rate).

An essential measurement

This measurement of the CEC and its saturation rate are essential data as they allow one to understand the chemical fertility of a soil, as well as the quantities of fertilizers and amendments needed for the mineral rebalancing of soils. Knowing the CEC helps to avoid applying more than the soil can retain, which would only result in losses through leaching. A low CEC requires reduced and frequent inputs.

18.2.2 Laboratory evaluation method

The most common method for determining the CEC is the ammonium acetate method (Metson, 1956). In this reagent, it is the ammonium (NH4+) that, thanks to its strong electric charges, displaces and replaces the adsorbed cations (Ca^{++}, Mg^{++}, K^{+} et Na^{+}) on the clay-humus complex. This method, reliable in soils around neutral pH, introduces biases in acidic soils by artificially raising their pH, and especially in alkaline soils by dissolving calcium carbonates. Indeed, ammonium acetate at pH 7 is "much more acidic" than the carbonates in a soil at pH 8 or 8.5. These carbonates are partially dissolved by the reagent (it is the unneutralized acetic acid that reacts), leading to the inclusion of calcium in the sum of the cations, which was not adsorbed on the CEC at all. This explains why, in saturated alkaline soils, the sum of the cations can represent well over 100% of the CEC (sometimes up to 200 or 250% of the CEC), which obviously does not accurately represent the actual occupancy of the CEC by the cations and their distribution.

A more reliable method in these extreme situations (acidic and basic) is the one based on cobalt hexamine buffered to the soil's pH. Unfortunately, it is adopted by only a few laboratories; the others prefer to retain the comparative value of their historical databases founded on the Metson method.

Saturation beyond 100%

The schematic representation of a CEC (cation exchange capacity) saturated beyond 100% by calcium and 0% occupied by H^{+} ions in very calcareous soils remains theoretical. Even in highly calcareous soils, negative charges may still be available and can carry H^{+} ions. These negative charges, occupied by hydronium ions, are mainly found on organic matter. The reality is more complex, and an approximate value of 0.5% hydrogen ions per unit of organic matter corresponds to the average reality (Mikhail, 2017).

18.2.3 Water pH KCl/CaCl pH

- **Analysis of water pH**

Measuring soil water pH involves placing distilled water in a soil sample (typically in a ratio of 2.5:1), which represents the soil solution in equilibrium with the sample. This pH reflects the soil solution's acidity or alkalinity as it interacts with plant roots. Soil possesses buffering capacity that moderates abrupt pH changes, influenced by factors such as organic matter inputs, fertilizer applications, tillage practices, temperature, and humidity. A decline in pH during spring and summer results from increased biological activity generating organic acids, oxidation of ammonium molecules, incomplete utilization of fertilizers leaving H_3O^{+} ions, and root exudates. Conversely, winter often sees an increase in pH due to hydronium ion dilution and an alkalizing effect from bacterial ammonification or nitrate reduction. Consequently, pH values can fluctuate by 0.5 to 1 unit within a season, significant given that pH is logarithmically linked to hydronium ion concentration. A one-unit pH shift equals a tenfold increase or decrease in acidity; pH 5 is ten times more acidic than pH 6 and a hundred times more than pH 7.

pH levels in agricultural soils typically range from 3.5 to 8.5. Soil pH measurements represent an average and do not fully capture the remarkable heterogeneity among micro-aggregates, the surface layer, and underlying horizons. Such variability makes precision beyond one decimal point less meaningful. This variability explains how pH-sensitive crops like barley can thrive in seemingly unsuitable soils. Within the analyzed layer and deeper, soil volumes conducive to normal growth can persist.

pH values do not inherently correlate with calcareous status and therefore should not solely dictate the decision to apply amendments. The alkalinity or acidity stems from the anion attached to calcium: H_2CO_3 (weak carbonic acid = $CaCO_3$ = pH 10),or H_2SO_4 (strong sulfuric acid = $CaSO_4$ = pH 6.5). Furthermore, pH can significantly fluctuate due to other cations like Mg, K, or Na, and therefore cannot automatically correlate with saturation levels.

■ Analysis of KCl/CaCl pH

The water pH measurement does not account for all acidic ions H^+ and Al^{+++} fixed on the CEC (cation exchange capacity). To assess their significance, a solution of KCl (potassium chloride) is used to extract these ions, which have weaker electrical attachments than potassium (K^+). The pH of KCl is always lower than water pH, with differences ranging from 0.5 to 1.5 point

■ pH, mineral availability and toxicity

pH strongly influences the availability and toxicity of major, secondary, and trace minerals. Here are the noteworthy pH values to remember:

- pH < 5.2 - 5.5: Toxicities may appear due to excessive solubilization of iron, manganese, and especially aluminum.
- pH 6.3 - 6.8: Ideal pH range where the best compromise for mineral availability is achieved.
- pH > 7.5: All minerals become less available, except for molybdenum, which experiences increased bioavailability.

18.2.4 Organic matter

Soil carbon levels are subject to fertilization management, climate regime, type of tillage and plant species cultivated.

■ Calculating soil organic matter content

The organic matter (OM) content reported in soil analyses is not a direct measurement but rather a calculated result based on measured organic carbon. Most European laboratories use a factor of 1.72 (Organic Carbon × 1.72 = OM), although many researchers argue that a factor of 2 would be more suitable for cultivated soils. This discrepancy can lead to significant variations in OM content reported by French and foreign laboratories. Referring to measured carbon content is more universal and comparable. In addition to these different conversion factors, the Loss on Ignition method is implemented differently across laboratories, with calcination temperatures ranging from 350 to 1100°C and durations from 2 to 32 hours. None of these methods is perfectly suited to all situations.

Measured organic carbon naturally excludes carbon integrated into carbonate rocks ($CaCO_3$), which is considered inorganic, despite its biological origin through calcium carbonation. However, certain parent rocks may also contain fossil organic materials, such as oil shale, potentially leading to overestimations of humus content.

Carbon levels are measured at the depth of sample collection. To account for soil management practices and changes in OM content due to tillage (e.g., no-till, reduced tillage, conventional tillage) and carbon storage, it would be important to use the Cumulative Humic Index (CHI), which involves summing up the measured contents over a depth of 60 centimeters (Baize, 1988).

It is common to distinguish different fractions of organic matter as follows:

- Particulate Organic Matter (POM) (> 50 μm), which are variably decomposed, often with a C/N ratio above 15, very low CEC (cation exchange capacity), and a short half-life (six months to ten years).
- Dissolved Organic Matter (DOM) (< 50 μm), which are more evolved, stable, and have a much longer turnover (> 100 years).

■ Increase in organic matter content as a deceptive indicator

There are numerous situations where an increase in soil organic matter content isn't the positive news expected, but rather a sign of blocked humification-mineralization processes. These unexpected effects commonly occur when:

- Acidifying organic materials are added (such as coniferous forest litter, slurries, or digestates from anaerobic digestion). This alters the fungi-to-bacteria ratio, slows down humification, and inhibits mineralization. Many recognizable plant debris remains (which do not contribute to CEC). This can lead to the paradoxical accumulation of poorly evolved acidic organic matter on calcareous soil and subsoil (e.g., in the Jura region).

- Calcic humus formation in high-limestone environments. Humus becomes inactive and no longer participates in normal carbon turnover. This situation is often exacerbated when transitioning from tillage to no-till practices. Mechanical impacts from tools like plows, harrows, or discs cease, and the limestone aggregates enclosing humus remain intact. While increased humus levels may be credited to no-till practices, they are not necessarily beneficial.

A reliable indicator of these described dysfunctions is the change in CEC (cation exchange capacity). When organic materials have properly evolved, each unit of organic matter generates a CEC of 2 to 3 meq/100 g. Deducting the CEC attributed to clay, finding values below 2 meq/OM should raise concerns and prompt investigation into the underlying issues rather than celebrating high raw SOM values.

18.3 Conclusion

An increase in organic matter content without a parallel evolution in Cation Exchange Capacity (CEC) should raise suspicions about the quality of biochemical processes occurring in the soil. A genuinely favorable evolution in organic matter content is typically accompanied by a simultaneous overall increase in CEC.

19 • FERTILIZATION

As long as we do not recycle all organic waste, animal and human excrement included, and as long as we do not return to a hunter-gatherer-fisher state where everything consumed, including their own remains, is left on site, and as long as we continue to cultivate plants that do not belong to the botanical families that thrived in the soils before cultivation, and as long as we aim for high yields on territories that were once heaths, wastelands, or sheep rangelands... this hope for total self-sufficiency will remain an illusion. Physical self-fertility (porosity, root growth) is possible, but chemical/mineral self-fertility is a dream. That is why fertilization remains a cornerstone of soil fertility maintenance.

The majority of fertilization guidelines presented are derived from agronomic research generated by the work of generations of researchers and agronomists: from Justus Liebig, Georg Liebscher, Gabriel Bertrand, André Voisin, William Albrecht (Missouri, USA), and finally Neal Kinsey, a contemporary American agronomist. There are some rules that structure this agronomic approach, which we present in this chapter. Adhering to these rules is the first step toward exerting a strong influence on soil porosity, water retention capacity, soil microbiology, and the nutrition and health of crops.

19.1 Basic principles

19.1.1 The Law of the Optimum:

Until now, everyone has remembered Liebig's "Law of the Minimum" (1850) and the famous barrel analogy with staves of different heights. Yield is linked to the level of the limiting element. However, we have forgotten the other law, which, though less known, is equally important. This is Georg Liebscher's **"Law of the Optimum"** (1895) sometimes referred to as the "Law of the Maximum," which establishes that beyond a certain level or ratio, an element can interfere so strongly that one or more others fail to play their role.

If there is too much calcium, then the availability of phosphorus, zinc, magnesium, iron, potassium, and manganese will be reduced. Excess nitrogen will hinder the absorption of boron, potassium, and copper. An excess of phosphorus will result in poor absorption of iron, calcium, potassium, copper, and zinc (Mulder's chart documents all these nutrient interactions well). In each of these situations of excess and induced deficiencies, diseases such as fusarium wilt, pythium, downy mildew, as well as physiological disorders like flower abortion, wilting, and hollow heart of beets, can appear, **without necessarily having very low levels of these elements in the soil.**

The almost automatic use of NPK ternary fertilizers sometimes brings about this "Law of the Maximum." Using N, P, and K according to the needs of plants, without considering soil reserves, is a risky endeavor. Moreover, the geological nature of the land itself can lead to unbalanced situations. This is evident in soils dominated by phosphorus, potassium, magnesium, calcium, or sodium. When we aim to grow plants in these soils that are not naturally adapted to them, it is crucial not to let physico-chemical imbalances persist, as they would disadvantage cultivated plants compared to endemic or invasive exotic weeds.

19.1.2 Rule of universality

All soils and all plants, cultivated in all climates, will respond to the principles we are about to discuss. Whether it's conifers, banana trees, wheat, corn, hops, azaleas, meadows, blueberries, etc., as long as the laws of physics, biology, and chemistry are respected, these crops will positively respond through yield and quality to soil rebalancing. The key understanding is that certain plants tolerate excesses better (such as barley with sodium and conifers with acidity, for example). However, if these plants are placed in a restored soil that, for instance, produces the most abundant and robust wheat crops, they too will thrive. From tropical red soils to temperate brown soils, from Moroccan sands to humic soils in northern Europe, from deep loams of Brie to the rich alluvial plains of the Nile, all will respond favorably to this approach and to cation rebalancing, provided water is not the limiting factor. When water ceases to be limiting, even mediocre soils can achieve their highest yields.

19.1.3 Choosing the right forms of fertilizers and integrating all their constituent elements

The functions of a fertilizing input are always multiple within the soil. The fertilization methods we advocate must take these various factors into account. For instance, when adding calcium, one must consider whether to use gypsum, ground limestone, or dolomite. Each of these forms has different effects and brings along companions that may not always be welcome. Calcium or magnesium have traditionally been seen as "neutralizers of acidity," but as we will see, they primarily impact the physical characteristics of the soil.

19.1.4 The choice of amendments is not dictated by pH

It is not the acidity or alkalinity that should dictate the use of an amendment! We can have alkaline soils that lack calcium (due to an excess of magnesium or potassium) and others that are acidic with already excessive calcium levels according to desirable ratios.

19.1.5 Feeding the soil before feeding the plant

- **Activating soil functions is the priority**

It has become increasingly common to consider that Ca, Mg, P, N, S, and K should be applied directly to the soil, while trace elements should be applied on foliage : Fe, Mn, Cu, Zn, B, and Mo. Foliar application is thought to be ten times more effective than soil application in terms of the percentage absorbed by the plant in the year of application. However, this overlooks the essential role that trace elements also play in the soil. For instance, zinc is a cofactor in over 300 known metalloenzymes across six enzyme categories: oxidoreductases, transferases, hydrolases, lyases, isomerases, and ligases. To support maximum efficiency of bacteria, fungi, decomposers, and humus builders, all these soil workers, artisans of fertility, need trace elements to activate their enzymes. These enzymes are crucial for a little-known facet of soil fertility. It is through enzymes, including free enzymes, that soils express their productivity and health, or conversely, their "fatigue." Enzymes act as catalysts for biological reactions, capable of multiplying the magnitude and speed of these reactions by a factor of 10, 100, 1,000, and sometimes much more. Slow processes of rock degradation, for example, can shift to an annual scale instead of taking centuries. Reclaiming the enzymatic capital and biodiversity is a must, and this minimally requires the presence of sufficient trace elements.

- **Limitations of foliar applications**

For plant nutrition, foliar feeding is possible but should be secondary. Regularly sampling leaves every fifteen days, sending them to a lab, and waiting to know what needs to be sprayed is neither reassuring nor sustainable. This approach might be validated for a few specialized soilless crops that, by definition, cannot benefit from the rhizosphere since it does not exist. Under such conditions, there is no humus to provide a balanced supply of nutrients as it mineralizes.

In open-field situations, it is the soil that must provide the bulk of the nutrition. **Both major and minor elements should be applied directly to**

the soil. Foliar applications should be technically nothing more than the "pinch of salt" needed for one reason or another. Managing inputs reactively based on foliar analyses is a poor practice, essentially a firefighting operation. These fertilization methods implicitly assume that the soil-root-plant chain is broken. This does not encourage the restoration of a virtuous functioning soil.

19.2 CEC: the first order of fertilization

The distribution of cations on the Cation Exchange Capacity (CEC) is the primary lever farmers have to revitalize their soils. This approach allows for controlling the behavior of different soil mineral fractions (sand, silt, clay). The proper behavior of these fractions leads to an optimal distribution of air and water within the soil. This is a prerequisite, before any other actions, to create an environment conducive to beneficial microbiology.

19.2.1 The ideal soil retains 25% air and 25% water

William Albrecht occasionally expressed the opinion that if he had to choose an ideal soil from a physical standpoint, it would be pure humus. When he began his career, he analyzed tens of thousands of soil and humus samples. His conclusion was that the most productive soils were those with calcareous humus, or mull. The humic A horizon has the following physical characteristics (see Figure 19.1). This is the goal we should aim for in all soils.

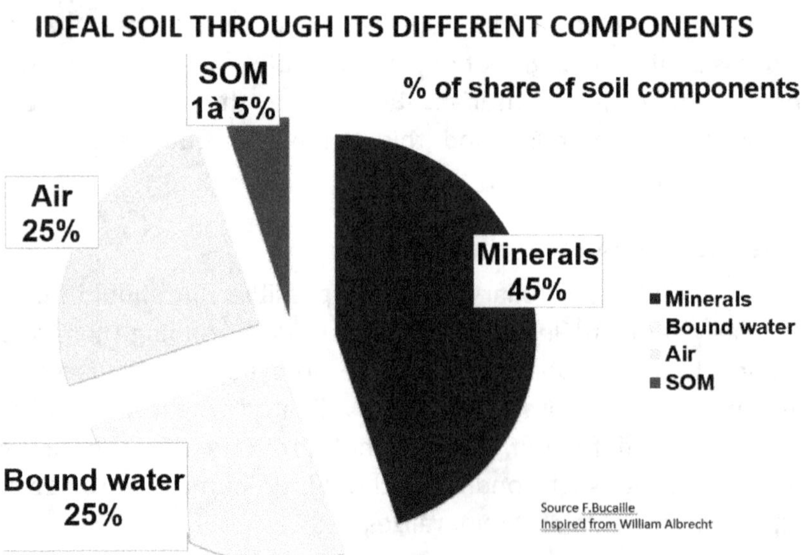

Figure 19.1 – The ideal soil in its mineral and organic components (% of soil fractions)

The ideal soil allocates 25% of its volume to water storage at field capacity. This is substantial and implies that the level of microporosity is at its maximum. Storing such an amount of water without causing hydromorphy means that this water is retained solely by weak electrical forces, known as van der Waals forces. This water is held on all the walls of capillaries, biopores, cavities, and also inside the clay particles. This characteristic is essential for storing water in winter without leaching and without causing asphyxiation, ensuring it can be released in the summer. The remaining volume will be occupied by air at 25%.

19.2.2 Controlling soil porosity through CEC

■ Two levers for an ideal soil

These distributions of the gaseous and liquid phases (25%/25%) are often reproduced in agronomy texts, but rarely does anyone explain how to achieve them. Similarly, we have all seen schematic representations of the perfect seedbed: a deeply fissured, porous soil with biopores, fine soil around the seed, and small clods on the surface to prevent crusting. These diagrams are a delight for graphic designers and a nightmare for farmers when it comes to actual implementation in the fields.

There are two levers to activate to reach this objective: one is biological activity, including roots, macrofauna, and microflora. The other is the action of minerals. It's not about choosing one over the other; both are necessary! However, the balance of minerals and cations is where it all begins. This is the first lever to allow a favorable microbiology to establish itself so that crops and cover crops can produce their maximum effect. Otherwise, efforts may be wasted, and time lost, with the hope that results will eventually come. For life to establish itself, basic needs must be met: shelter and sustenance with water and oxygen. This requires managing the soil's physical properties!

■ A lever on clay behavior

What is the relationship between soil porosity, water retention, and cations adsorbed on clays and humus? In reality, through the charge of the Cation Exchange Capacity (CEC), we can influence the behavior of clays. Depending on the distribution of cations, clays will be more or less cohesive or, conversely, more or less "loose." To ensure sufficient air and adequate water retention, the base saturation and the proportions of cations expressed in meq on the CEC should be as follows (see figure). In these proportions, calcium, magnesium, potassium, and sodium do not compete and can be absorbed by plants as needed. However, the most significant impact of these cations being in the correct proportions is on the physical structure of the soil itself.

■ The calcium-magnesium ratio: a tuning knob

To manage clay behavior, the first parameter to consider is the calcium-magnesium ratio on the Cation Exchange Capacity (CEC). Magnesium tends to act like a "glue" for clays, retaining water. In contrast, calcium provides flexibility and aeration to the soil. The calcium-to-magnesium ratio is a tuning knob between these two opposing soil tendencies. The best air/water ratio is achieved with a 68:12 (Ca/Mg) ratio. Altogether, calcium and magnesium should occupy 80% of the CEC (in meq).

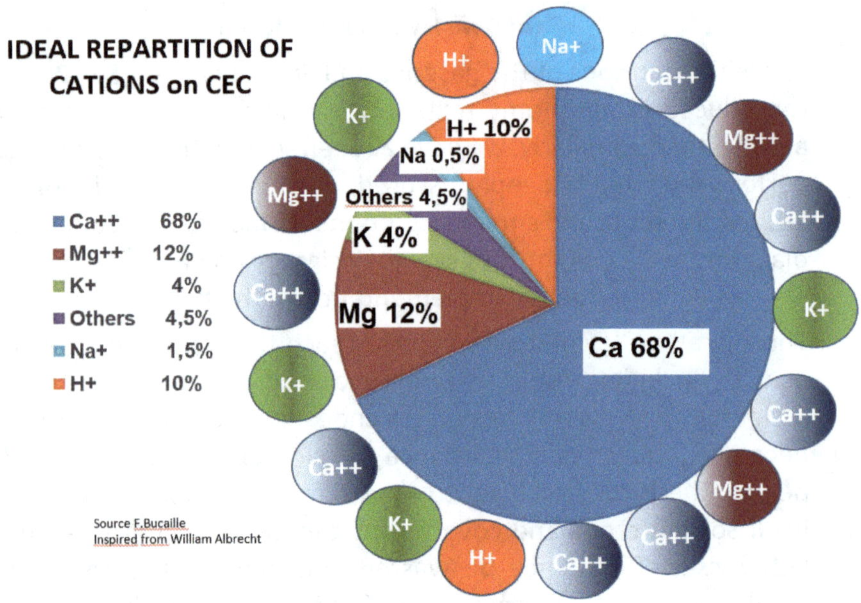

Figure 19.2 – Ideal cation distribution on the CEC (according to William Albrecht)

Increasing the calcium share in this ratio results in more air occupying the pores (ventilation/aeration). Increasing the magnesium share, conversely, results in more space being occupied by water (moisture/cohesion).

When considering the notion of heavy or light soils, we prefer to refer to the soil's CEC, which, in our view, better reflects the presence of true clays than granulometry through a sedimentation test conducted in a laboratory.

■ Adjusting the ratio according to the situation

The ideal ratio of 68:12 should be adjusted according to specific situations, ensuring that the two elements always occupy 80% of the CEC. When a soil has a very high clay content with significant CEC values, it is advantageous to maintain a relatively low percentage of magnesium, around 10%, to reduce soil cohesiveness and allow more air to enter. Conversely, in very light soils with low CEC (50 meq/kg), it is beneficial to help the soil retain and store as much water as possible, aiming for a magnesium level of 20%. Given that calcium and magnesium together should represent 80% of the CEC, we have established a scale of desirable values as follows:

Tableau 19.1 – Ideal calcium/magnesium ratio according to soil CEC (in meq/100 g)

CEC/100 g	Ä 5.40	5.41 - 5.7	5.71 - 6	6.01 - 6.3	6.31 - 6.6	6.61 - 7	7.01 - 7.7	7.71 - 8.3	8.31 - 19	19.01 - 45	> 45.01
Ratio %Ca/%Mg	60/20	61/19	62/18	63/17	64/16	65/15	66/14	67/13	68/12	69/11	70/10

Adjusting for ecological climax

The Ca/Mg ratio is a refined means to maximize the physical fertility of soils. It also allows for cultivation in greater harmony with the original climax cycle: storing water when necessary in winter without causing hydromorphy and releasing it in spring and summer. While magnesium and calcium are considered secondary elements for plants compared to nitrogen, phosphorus, and potassium, they are the most important for living soils due to their significant influence on the quality of the habitat for microorganisms and roots.

19.2.3 Qualities of the main CEC cations

■ Only calcium and magnesium have a structural effect

Only calcium and magnesium can play a structural role in soil. The other two major cations encountered, potassium and sodium, are monovalent (carry only a single positive charge). Therefore, they act as "dispersants" when present in excessive quantities in soil. Another characteristic is the amount of water these atoms aggregate around them. It is these physical characteristics that give them the effects we are describing on soil structure.

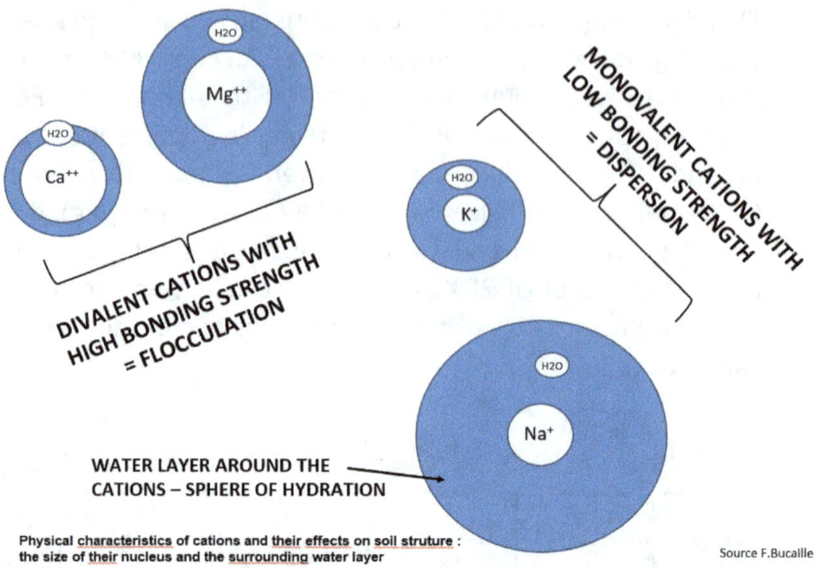

Figure 19.3 – Physical characteristics of cations: reasons for their effects on soil structure: the size of their nucleus and the surrounding water layer

■ Properties of cations present on the CEC

- **Mg^{++}** flocs soils (provides structural stability) due to its double valence (two positive electrical attachments). It efficiently binds clays and humus together and helps soil retain water because it is surrounded by a significant water layer. This characteristic also prevents dust and sand loss through wind erosion as it "glues" these elements together. However, in excess, especially with clay present, soils can become so compact and impermeable that hydromorphy can occur. The extreme soil compaction can even lead to plant root system atrophy. Paradoxically, magnesium deficiencies may appear in such cases!
- **Ca^{++}** also flocs soils effectively due to its double valence and stabilizes the clay-humus complex. However, its dominant salt in soils, calcium carbonate ($CaCO3$), is not hydrated at all. This means its flocculating action dominates, and its ability to retain water in the soil is very low, thereby reserving maximum volume for air
- **K^{+}** is monovalent and therefore cannot stabilize the soil; it has a dispersing effect. Additionally, it has a significant water envelope: its excess causes soils to compact and become very hard. Certain clays (illites) can be irreversibly damaged.
- **Na^{+}** is also monovalent but with an even larger water envelope than K^{+}. Therefore, it is more soil-dispersing than potassium. Sodium is not

routinely measured in soil analyses; it is only measured in situations where excess is a concern. Beyond 5% sodium on the CEC, negative effects appear on structural stability and electrical conductivity. However, there is also a too-low threshold: below 0.5% of the CEC or < 10 ppm, there is too little sodium to support a highly active micro and macrofauna, including earthworms. All animals, including the smallest ones, require sodium for maintaining osmotic pressure, proper muscle and nerve function.

20 • MANAGEMENT OF AMENDMENTS AND FERTILIZERS

The subject of the reflection proposed in this work concerns only those elements that will have a lasting influence on the soil and should be considered as basal fertilizers or even amendments, although some may be classified as fertilizers under regulations. The issue of nitrogen will also be addressed here, albeit briefly.

20.1 Calcium (Ca) management prevails to rebalance CEC

The rebalancing of calcium on the CEC according to the ratios we have discussed takes precedence over all other cation rebalances. Even for the aggressiveness of pathogens, insects, and aggressive weeds, the absolute and relative levels between calcium and magnesium are decisive. Additionally, calcium strengthens the skins of all vegetables and fruits by causing pectin gelation, ensuring long-term shelf life and exceptional organoleptic qualities. Human perception of sweetness depends not only on sugar content but also on calcium levels, as taste buds have sugar receptors activated only in the presence of calcium (Calcium is a neuro-transmitter). For these reasons, calcium should be the first topic addressed in any soil revitalization program. The discussion of calcium should focus on its intrinsic importance rather than its effects on pH.

20.1.1 Balanced pH is the result of a living soil

We have established as a guiding principle that pH and acidity should not be the starting point for informed agronomy. While pH is important, in our approach, it will merely reflect the recalibration based on the ratios discussed in the previous chapter. The balanced Ca/Mg ratio (68/12) is the prerequisite for a healthy balance between bacteria and fungi. It is the primary lever to reduce climactic disruption, enabling active vegetation during summer by providing fungi with the water and oxygen they need. Under these conditions, fungi can occupy their ecological niche and fulfill their functions without being overwhelmed by bacteria. When this balance is restored, pH naturally tends towards more moderate values for the soil. Life creates conditions conducive to more life! Regarding fungi, it's important to note that an optimal soil pH ranges from 6 to 6.5. It is within this range that elements are generally most available (except for molybdenum).

20.1.2 Calcium requirements take precedence over pH when deciding on an amendment

Therefore, it can happen that the most urgent measure to take is to apply calcium, even when pH levels are high. Conversely, in some acidic soils (especially those with gypsum), it may be urgent to refrain from any liming.

The primary function of liming is to supply calcium ions, not to increase pH. As illustrated by the following example, high pH levels and liming can coexist beneficially. The clayey soils of the Limagne area often have a high magnesium content, ranging from 20% to 25% on the CEC (cation exchange capacity). These soils are very heavy, sticky, compact, and difficult to work with. Their pH is high, between 7.8 and 8.2. Yet, to restore the structural functionality of the soil, calcium amendments produce very good and rapid results. Indeed, calcium inputs are effective in rebalancing the distribution of cations on the CEC (by displacing some of the magnesium present). Given that magnesium is 1.4 times more alkaline than calcium, this tends to lower the pH. However, the benefit is not merely the pH decrease itself. More importantly, soil porosity improves immediately. The soil shows a reduction in stickiness, becomes easier to work with, and agronomic results improve significantly. This approach, which runs counter to conventional wisdom, was successfully experimented with after reading the works of American researcher William Albrecht and in collaboration with American agronomist Neal Kinsey.

In soils with a high abundance of magnesium (above 12% on the CEC in clayey soils), magnesium is responsible for the cohesion of clays and the heavy structure. The strong cohesion of clays leads to the rapid atrophy of root hairs, and the immediate root environment becomes depleted in magnesium, which is not very mobile. Magnesium deficiencies are commonly observed in these soils despite their high magnesium content. Farmers, unaware of this phenomenon and guided by their technicians, often mistakenly think the issue is the form of magnesium in the soil. This misconception only exacerbates the situation. In such cases, it is crucial to "rescue" the crop by applying magnesium strictly in foliar form based on foliar analyses confirming the deficiency. Simultaneously, remedying the situation by applying calcium carbonate or preferably gypsum is essential.

20.1.3 Scenarios for managing calcium amendments

Prior to attempting to manage the other cations on the Cation Exchange Capacity (CEC), it is first necessary to seek to rebalance calcium. However, the type of calcium amendment chosen will depend on the status of the other cations. We have illustrated various scenarios and the preferred corrective amendments based on the situations encountered in the form of a summary diagram. The starting point of the diagram and the reasoning is calcium **Ca**$^{++}$. Subsequently, reference is made to the respective levels of the other cations on the CEC to determine the different types of corrective amendments that can be applied to achieve their best possible distribution on the CEC.

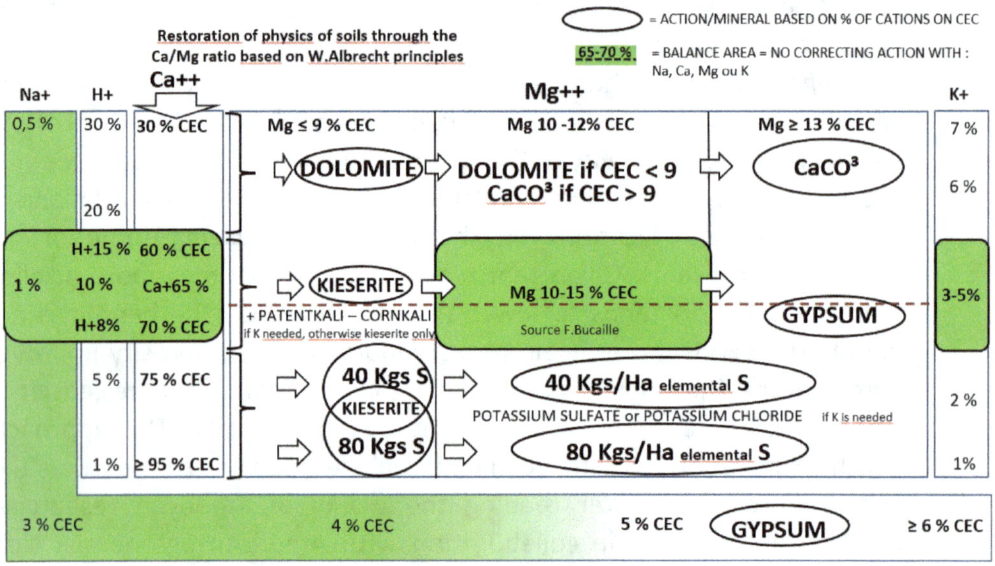

Figure 20.1 – Diagram for managing calcium amendments based on the distribution of cations on the CEC. Revitalization of soil physical fertility through action on the Ca/Mg ratio, based on the works of W. Albrecht

- **Calcium deficiency (lack of soil porosity)**

- **Dolomite**. This amendment is suitable if the Cation Exchange Capacity (CEC) is deficient in both calcium and magnesium (< 9%).
- **Calcium carbonate**. Consider this option if only calcium is lacking.
- **Gypsum.** This calcium source is preferred in cases where soils have excess magnesium (> 13%). However, gypsum works well to control excess magnesium only when calcium saturation reaches 60%, which can be initially achieved with calcium carbonate if necessary. For crops that require low pH, such as azaleas, blueberries, rhododendrons, and

potatoes, we meet the soil calcium rebalancing needs using gypsum, which does not raise pH.

- **Lime** is not included in the diagram because burnt lime products (quicklime or hydrated lime) must be absolutely avoided! They are biocides (burn all living tissues). While they generally yield good crop yields, they come at the cost of depleting organic matter, significant loss of macrofauna, and sometimes the breakdown of certain clays. They should be strictly avoided..

■ Neutralizing excess calcium (low water holding capacity, chlorosis, mineral deficiencies, etc.)

When there is an excess of calcium, the initial consequences include reducing the soil's water storage capacity, creating a structure that easily turns into dust, and causing deficiencies in magnesium and potassium, iron chlorosis, and various induced deficiencies. In this situation, we emphasize two phases of neutralization. The first relies on amendments such as patentkali, kieserite, and elemental sulfur. The second relies on soil fungi.

□ First phase: mineral chemical approach

Patentkali: This amendment is used when there is a simultaneous need to supply magnesium and potassium.

Kieserite: This amendment is preferred if only magnesium is lacking.

Sulfur: Applications of patentkali and kieserite often benefit from being complemented or preceded by elemental sulfur (S) application, which neutralizes calcium in its interactions with trace elements. The mechanism works as follows: after oxidation of sulfur by soil sulfobacteria to SO_2, SO_3, and then SO_4, excess calcium in the soil solution crystallizes into calcium sulfate salts (gypsum). This process from native sulfur is slow, which is advantageous as it will accompany the soil throughout a season, gradually neutralizing calcium as it dissolves into the soil solution. The chemical transformation of the gypsum formed will not lead to calcite crystal formation. It will concentrate less around the roots, thereby reducing the risk of iron chlorosis. This method effectively reduces the amount of calcium in the soil solution.

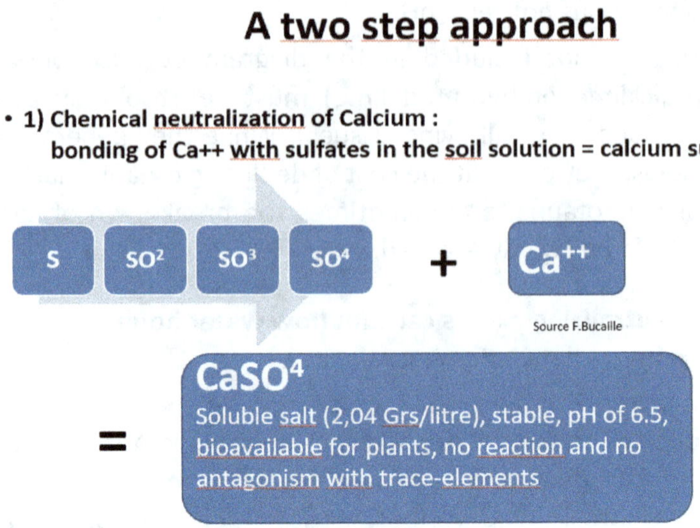

Figure 20.2 – Chemical neutralization of calcium by sulfates

□ Second phase: biological pathway through soil fungi

Soil fungi produce organic acids, including oxalic acid abundantly, which neutralize calcium, as well as many other minerals in excess in the soil solution (iron, manganese...) by crystallizing them into oxalates. These oxalates are completely insoluble, slightly acidic in pH, yet 100% bioavailable. These forms of mineral storage are the most effective and the only ones that can prevent surface acidification risks in no-till and conservation tillage systems (see diagram below).

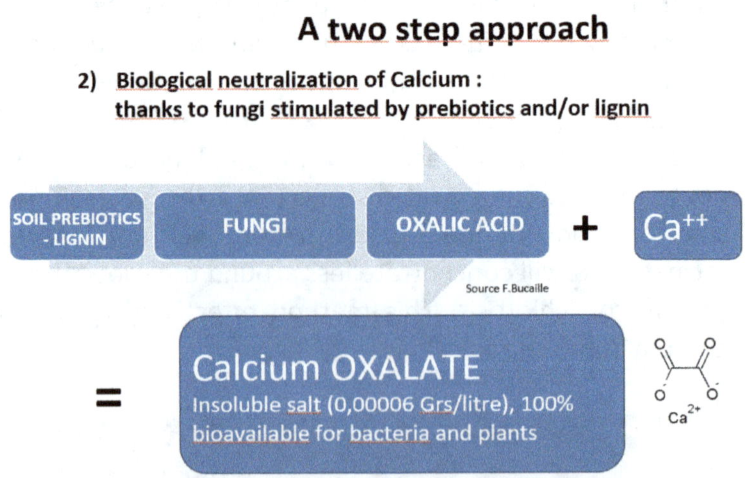

Figure 20.3 – Calcium (or iron, or magnesium) neutralization in the form of oxalate. Recreating natural, sustainable conditions (lasting several years) favorable to calcium neutralization through fungi stimulated by soil prebiotics.

Table 20.1 – Effects of dominant forms of calcium on calcium dynamics

Soil condition	Dominant form of calcium in solution	Solubility of Ca in g/liter	Decalcifying and chlorotic effect
Dysfunctional soil	Calcium Carbonate Calcium Bicarbonate	0.014 g/liter 16.6 g/liter	++ +++++
Corrected SO_4 soil	Calcium Sulfate	2.04 g/liter	=
Living soil/fungi	Calcium Oxalate	0.00006 g/liter	---

Bioavailable forms of calcium

These two forms of calcium (sulfate and oxalate), derived from the two neutralization pathways we have just presented, are of great interest for all perennial crops sensitive to chlorosis, as well as for crops such as potatoes. Indeed, potatoes absolutely need calcium to strengthen their periderm, but at the same time should not come into contact with CaCO3, which promotes the development of Streptomyces scabies (common scab). Calcium oxalate is completely assimilable without the disadvantage of calcium found abundantly in soils where fungal flora has almost disappeared.

20.1.4 Calcium also interferes with many other trace elements

Table 20.2 – Possible interference of calcium with various elements. For 1,000 kg of free calcium in the soil in the form of $CaCO_3$, the elements iron, manganese, copper and zinc will be rendered unavailable to plants in the following proportions:

Salt of trace element concerned	Quantity of trace element salts neutralized per 1000 kg of calcium
Iron sulfate	80 kg
Manganese sulfate	40 kg
Copper sulfate	8 kg
Zinc sulfate	15 kg

20.2 Consideration of other minerals

20.2.1 Magnesium (Mg)

When there's **too much** magnesium on the CEC, as we've already seen, the soil becomes very "sticky" in wet weather and hard as concrete when it dries out. Magnesium then interferes negatively with the availability of phosphorus and potassium. An excess of magnesium can even lead to magnesium deficiency in plants, as an indirect effect of degraded soil structure (see the case of Limagne soils above). This excess can also be controlled through the use of sulfur, which combines with magnesium to form a soluble salt that can be removed.

Another little-known effect of magnesium is its impact on nitrogen dynamics and uptake, through direct interference and degradation of microporosity and root density. The greater the excess, the lower the nitrogen utilization coefficient.

Table 20.3 – Loss of nitrogen efficiency per additional point of excess magnesium on CEC (Source Neal Kinsey)

Exceeding/Calculated ideal Mg ratio	Additional N consumption for the same yield
= Mg/CEC	=
Mg/CEC + 1	+18%
Mg/CEC + 3	+32%
Mg/CEC + 5	+45%

20.2.2 Sulfur (S)

Sulfur is as essential as nitrogen, contributing to the construction of similar organic molecules such as proteins, enzymes, vitamins, and phytoalexins, in an S:N ratio of about 1:15. Its importance was overlooked because, until recently, it was adequately supplied by atmospheric pollution (during peak sulfur pollution in Western Europe, approximately 50 kg of S/ha/year would be deposited). Sulfur is highly mobile in soils, like other anions such as nitrates, borates, and molybdates. Its losses and efficiency are directly linked to the Ca/Mg ratio (averaging 68/12), following a model similar to that of nitrogen (refer to the nitrogen efficiency loss table). When present in excess, sulfur can antagonize other anions like phosphorus.

In our agronomic approach, sulfur is used as a fertilizer in acidic to neutral soils. It is recommended to apply it at doses strictly necessary for the crop. In alkaline soils where an excess of cations (such as calcium, magnesium, potassium, sodium) may be present, sulfur is used marginally as a fertilizer and primarily as an amendment. The purpose of sulfur application in these soils is to modify the physico-chemical characteristics by neutralizing the effects of excessive cations on the soil's physical behavior (dispersive or cohesive effects). Additionally, sulfur has complexing effects on other minerals.

20.2.3 Nitrogen (N)

Nitrogen, along with carbon, is central to the molecules of life. Plants extensively use nitrogen to construct their amino acids, vitamins, and enzymes. Nitrate forms (NO_3^-) primarily contribute to vegetative growth, while ammonium forms (NH_4^+) support fruiting. Unlike the carbon cycle where plants extract carbon from atmospheric CO2, which is present in very small quantities (0.0415% of the atmosphere), nitrogen fixation from atmospheric nitrogen (78% of the atmosphere) is solely performed by bacteria. All consumed nitrogen thus originates from either industrial processes (Haber-Bosch process) or bacterial activity. Only free-living bacteria (*azotobacter, azospirillum,* etc.), symbiotic bacteria (rhizobium), and algae have the ability to convert atmospheric dinitrogen (N_2) into forms assimilable by plants (NH_2/NH_4). They utilize an enzyme called nitrogenase, with its core reactor containing molybdenum ions, complemented by cobalt. Hence, the importance of having these in all soils. To capture nitrogen from the air, bacteria require air and moisture. Therefore, Ca/Mg ratios are crucial, even for legumes, which are only partially self-sufficient in nitrogen, ranging from 50% to 70%.

Nitrogen deficiencies due to poor absorption occur under the following conditions:

- When there is less than 60% calcium around the roots.
- When the pH is high with less than 12% magnesium.
- When magnesium levels are too high (>12% in heavy soils). In heavy soils with an Mg/CEC ratio >18%, the nitrogen requirement can be more than 50% higher due to soil compaction and poor aeration.

■ **Possible causes of N-NO_3 deficiency**

- Excessive chloride (hence the importance of using potassium chloride cautiously).
- Poor Ca/Mg ratio.

■ **Origins of N-NH_4 deficiency**

- Elevated levels of magnesium leading to competition of NH+ with Mg++ ;
- Excess potassium.
- Soil temperatures are too high (>25°C).
- Absence of mycorrhizae: NH_4 is less mobile as it is fixed to clays, unlike nitrates. Fungal hyphae transport N-NH_4 over distances of 10 to 15 centimeters.

Note: For optimal plant health, the sap ideally should contain 75% NH_4 and 25 % NO_3.

■ **Excess nitrogen and disease sensitivity**

Excess nitrogen makes plants more susceptible to diseases, a sensitivity exacerbated by the poor assimilation of the following elements:

- Copper
- Manganese
- Potash
- Zinc
- Calcium.

Important: Excess nitrogen causes calcium to leach out (in the form of calcium nitrate, a very soluble salt) without having the same effect on magnesium. This degrades the Calcium/Magnesium ratio, causing the soil to harden and compact.

Synthetic nitrogen itself is not harmful to the soil. Only the nitrogen that is not used by crops and leaches away causes a loss of bases, including calcium, and therefore acts as a decalcifying agent. Additionally, it is acidifying by leaving behind hydronium ions H^+ generated previously by the oxidation of urea or ammonium nitrate into nitrates.

$$NH_4NO_3 + 2O_2 = \mathbf{2H^+} + 2NO_3 + H_2O$$

If all the nitrates are used, the equation balances out by the reorganization of nitrogen:

$$R\text{-}OH + NO_3^- + \mathbf{H^+} = R\text{-}NH_2 + 2O_2$$

Excess nitrogen fertilization also affects the fate of soil organic matter (SOM). For sugarcane, exceeding 100 kg N/ha triggers the process of mineralization and carbon loss in the soil (Almeida Assunção et al., 2018). It also appears that the nitrogen applied can trigger carbon losses in the soil, with a differentiated effect on the utilization of nitrogen itself depending on the forms used (Khalil et al., 2007). In a study conducted on six types of soils (acid sulfate, peaty, saline, non-calcareous, calcareous, terraces), carbon losses over 90 days ranged from 2.3% for acid/peaty soils to 10.6% for calcareous soils. Additionally, the study concluded that urea fertilizers seem more suitable for acid soils, while ammonium sulfate is better suited for calcareous soils.

20.2.4 Phosphorus (P)

Phosphorus is crucial for plants, aiding in energy transfer (ADP/ATP - Adenosine Diphosphate and Triphosphate) and cell division. It needs to be sufficiently available in soils. Phosphorus is vital for plant growth in cold and wet conditions, for better water utilization, and for early maturity.

Phosphorus behaves uniquely for an anion: unlike other anions (nitrates, sulfates, and molybdates), it is not leachable and is very immobile in soils. It readily forms strong chemical bonds with other minerals: with iron and aluminum in acidic soils, and with calcium in alkaline soils, making it unavailable to plants. To prevent these mechanisms, it is essential to first adjust the Ca/Mg ratios as discussed earlier, because phosphorus availability is closely linked to the neutralization of iron and aluminum ions, soil aeration, and good biological activity. These conditions mainly depend on the level of calcium and magnesium saturation in the CEC (Cation Exchange Capacity), with an average ratio of 68:12. Maintaining this balance allows for denser root growth by improving soil porosity, enabling roots to reach phosphorus. Since phosphorus is immobile, it will not move toward the roots as nitrates do; instead, roots must grow toward the phosphorus.

▪ Mycorrhizae enhance absorption

Active microbiology is essential for making phosphorus available to plants. Phosphorus absorption heavily depends on microbiology, particularly on mycorrhizae. These fungi have hyphae so fine (two to ten microns, meaning they are invisible to the naked eye) that they can penetrate spaces where roots cannot reach, exploring the tiniest crevices in the soil to extract, solubilize, and transport minerals. Mycorrhizae contribute significantly to soil stability through their hyphae and the production of glomalin, a glycoprotein with remarkable adhesive properties and longevity. This molecule, discovered in 1996 by Sarah Wright of the USDA, accounts for 30% of the organic matter in living soils. Phosphorus availability is also tied to the humus content in soils, which is why promoting humification processes enhances these humus stocks and their accessibility. Thus, activating the soil's microbiological processes is crucial for maintaining and improving phosphorus availability.

▪ Antagonisms

The antagonism with sodium (Na) can manifest when sodium levels are high (> 5 % of CEC). In such situations, phosphorus deficiencies may occur even if phosphorus levels are considered sufficient.

The zinc/phosphorus antagonism is so strong that specific ratios should be maintained to ensure proper balance. The recommended ratios (depending on the phosphorus measurement method) are as follows, without the total zinc exceeding 50 ppm:

- Olsen 1:10
- Bray 1 : 15
- Mehlich 3 1 : 18

When playing it safe means applying a limited dose of phosphorus

The relationships between phosphorus, plants, and mycorrhizae are intricately intertwined. It is essential to have enough phosphorus to ensure vigorous crop establishment and continue availability throughout the growing season, as its uptake is continuous. However, it is prudent not to exceed certain levels. Phosphorus levels should be low enough to trigger significant production of strigolactones, which are hormones that attract mycorrhizal hyphae to plant roots. Strigolactones play a crucial role in mycorrhizal colonization. After mycorrhizal spores germinate, the hyphae have seven to ten days to find a root that will supply carbohydrates; otherwise, they die from a lack of energy. During this critical period, the initial hypha reacts to the presence of strigolactones by forming multiple branches, increasing its chances of finding a plant root. Thus, mycorrhizae infect roots and become functional only at low to moderate phosphorus levels.

Proper phosphorus management, which initiates mycorrhization, is a key component in mitigating climatic shifts. The benefits provided by mycorrhizae extend beyond phosphorus supply, as they also improve water access for crops. By balancing phosphorus levels to encourage mycorrhizal activity, plants can achieve better growth and resilience.

Table 19.4 – Indication of high phosphorus thresholds (P2O5) not to be exceeded in soils in order to generate a beneficial mycorrhizal association for the plant.

Methods	Upper phosphorus (P2O5) limits not to be exceeded
Olsen (alkaline soils)	60 ppm
Bray 1	100 ppm
Mehlich 3 (acidic or neutral soils)	120 ppm

Mycorrhizae: Mining power and suppression of pathogens and weeds

Mycorrhizae, present in 95% of cultivated soils according to our analyses (only soils that are regularly fumigated, heat-treated, or chemically disinfected lack them), play a major role in nutrient transport. In addition to phosphorus, they efficiently transport nitrogen, copper, iron, magnesium, manganese, and water. Through their exudates, they foster a highly active "hyphosphere" of bacteria that mineralize humus and solubilize minerals. Mycorrhizae's extraction power, thanks to their hyphae that are two microns in diameter, enables them to compete with soil pathogens like Fusarium, Rhizoctonia, Pythium, Verticillium, and Phytophtora for nutrients. They also give mycorrhizal crops a decisive advantage over non-mycorrhizal weeds such as Chenopodiaceae, Amaranthaceae, Brassicaceae, and Polygonaceae. These families include common weeds like knotweed, bird's knotweed, field pennycress, wild mustard, shepherd's purse, etc. It is crucial to manage phosphorus fertilization carefully to ensure high productivity without increasing competition from soil pathogens and certain weeds. Maintaining phosphorus levels within recommended thresholds encourages beneficial mycorrhizal associations while suppressing undesirable plant species. Only non-mycorrhizal crops, such as beets and Brassicaceae (rapeseed, mustard, cabbage), require careful monitoring without exceeding the indicated phosphorus values.

20.2.5 Potassium (K)

Photosynthesis, transport and storage of carbohydrates, cold hardiness, tissue strengthening and lodging resistance, protein synthesis and improved drought resistance. Potash is important for plant metabolism! However, it is not a constituent of organic molecules, and is essentially present in intercellular spaces (not inside cells). This is why straw and manure leach potassium so easily, and why there can be enormous wastage when composts and manures are stored at the head of fields over winter.

The desired values are between 2 and 7% of CEC. As far as this element is concerned, the percentage of CEC can no longer be a reference when the exchange capacity is very low (< 5 meq/100 g). In this case, 100 ppm should be considered as a threshold value, with the possibility of a second or even third application during the season for demanding plants. When the CEC is very low (as low as 2 meq/100 g, for example), 4 or 5% per CEC would not provide sufficient quantities to satisfy the plant's simple needs.

■ Non-constitutive of plant tissues

Considering that potassium is not a constituent of plant tissues and primarily functions in osmotic balance, it is largely returned to the soil by crops at the end of their cycle through root exudates. For instance, rapeseed, known for its high potassium needs, may require 250 kg of potassium during its growth but will export only 10% of this amount. **The critical point is that the soil can mobilize and "lend" this quantity** to the crop. This mobilization depends on good soil porosity and significant root density. Plants harvested before maturity export all nutrients present in their tissues at that time. In such cases, nutrient needs and exports are nearly equivalent. This applies to crops like vegetables, potatoes, sugar beets, alfalfa, silage maize, and cover crops for biogas.

■ **Thresholds not to exceed**

It is desirable to maintain potassium levels at 5% for annual crops and 7% of the Cation Exchange Capacity (CEC) for perennial plants such as fruit trees and vines. However, exceeding these thresholds can lead to competition for the absorption of calcium and magnesium. The potassium/calcium antagonism is particularly crucial for plants with quality requirements related to appearance, storage, and taste, such as potatoes, fruits, and root vegetables. Calcium is essential for skin strength and pectin gelation.

Adjusting potassium to soil content

> **Potassium should be managed based on the actual soil content rather than solely on crop needs.** For instance, applying 300 kg of potassium per hectare to a potato crop without first checking soil potassium levels can be risky. Such excessive fertilization could raise potassium levels above 7% of the CEC, increasing susceptibility to shocks, skin diseases like scab, and storage-related disorders. Insufficiently absorbed calcium would fail to solidify the skin, compromising crop quality and storability.

20.2.6 Boron (B)

Boron has a number of essential functions, one of the most important of which is to transport calcium. Applying calcium to the soil or foliage, in the hope of finding it in fruit or leaves, is a waste of time and money if the soil is deficient in boron.

Boron is also vital for pollen fertility, fruit set, nitrogen efficiency (a boron-deficient soil will require more nitrogen through poor absorption), nodule formation in leguminous plants, and sugar transport in plants. Plants that assimilate and export large quantities of calcium (alfalfa, for example) and store a lot of sugar (beet, sugar cane, fruit trees, vines) require boron levels of 1.5 to 2 ppm. Other plants can cope with lower levels of 0.5 to 1 ppm.

■ **Challenges posed by boron**

- Unlike most other trace elements, which are catalysts, boron is "consumed" and not recycled, which means that a constant supply must be ensured. Like many anions, boron is leachable, especially in sandy soils. In this type of soil, it is sometimes useful to make several additions per year.
- It is one of the most toxic trace elements, hence the need for frequent low-dose supplements in case of deficiency. This toxicity is more pronounced in soils with low calcium levels. Boron should be at a ratio of 1:1,000 with calcium, without exceeding 4 ppm.

- Deficiencies are more frequent in soils with high potassium or calcium levels. A calcium content > 50% CEC and < 80% CEC reduces boron toxicity but optimizes absorption.
- More frequent deficiencies in soils with low humus content

■ **Inputs**

Soil applications should be made at a maximum rate of 2 kg/ha of elemental boron at a time (around 20 kg of borax or boron ethanolamine). Its borate form makes it leachable. It should therefore not be applied to the soil in autumn. Foliar applications should not exceed 200 g B/Ha.

20.2.7 Chlorides (Cl)

Chloride levels in soils can be naturally elevated. The extensive use of potassium chloride (KCl) fertilizer can potentially raise soil chloride levels to penalizing values, although occurrences above 1,000 ppm are rare. Chloride ions, contrary to what some rural legends claim, are not toxic to living organisms (we're not talking about chlorine gas Cl2 or sodium hypochlorite NaClO), but on the contrary contribute to the efficiency of photosynthesis, increase the size of leaf cells and improve nitrogen efficiency.

20.2.8 Cobalt (Co)

Cobalt is an important element in the health of animals that ingest plants. It is in fact the basic element of vitamin B12 (cobalamin), synthesized by rhizospheric bacteria that transfer it to the plants that host them. It is also an important element in symbiotic nitrogen fixation. In case of deficiency, apply a maximum of 250 g/ha of cobalt sulfate per pass, twice a year.

20.2.9 Copper (Cu)

Copper, although absorbed in very small quantities, is involved in photosynthesis, protein synthesis, N-NH_4 absorption and stem strength (along with manganese and potassium), as well as conservation.

Availability: the main antagonists to the absorption of this trace element are high levels of humus, sandy soils, high doses of nitrogen and heavy phosphate fertilization. Secondly, the ubiquitous calcium is involved in copper blockage. A level of 2 ppm is a very good value. Above 30 to 50 ppm, common in viticulture and arboriculture, particularly in organic farming (where values of up to 300 to 400 ppm are found), there is a marked toxicity: on soil decomposition fungi, on highly sensitive nitrogen fixing bacteria, on anecic fauna and above all on small epigeic surface earthworms.

Application: foliar applications of copper should not exceed one kilogram of copper metal per hectare per pass. It is best applied to the soil in the absence of vegetation, in granular form or as a spray. Soil application of copper sulfate can reach 40 kg/ha/year. Like zinc, copper is very stable in the soil and will not leach out. The value of copper should never exceed that of zinc.

20.2.10 Iron (Fe)

Iron is involved in chlorophyll function, energy transfer and various enzyme systems. Iron is mainly absorbed by the roots.

Availability: its availability is negatively influenced by high pH and a calcium-saturated base (ferric chlorosis). Excessive magnesium, potassium or sodium levels have the same effect. Highly acidic soils are not immune to iron deficiency, although toxicity is the most common situation in these soils. Sulfur plays an important role in improving iron availability. On average, the interference between calcium and iron is of this order: one tonne of calcium carbonate will block the equivalent of 70 kg of iron sulphate under the real conditions of a fertilization program.

Application: ferric sulfate should not be applied to the soil at doses exceeding 400 kg/ha of ferric sulfate.

Caution: if manganese levels are higher than iron levels in the soil, iron is oxidized in plant tissues and therefore becomes ineffective, even if leaf analysis reveals sufficient iron content! This relatively common problem needs to be solved at soil level first.

Liming with calcium carbonate or dolomite can also result in induced iron deficiency when the value was already low. The same applies to other trace elements. The finer the liming material, the higher the risk.

20.2.11 Manganese (Mn)

Manganese has been active since the earliest stages of plant life. It is one of the most essential trace elements for plant metabolism, especially when large quantities of glyphosate are used, as the latter creates powerful electrical bonds with it, making it unavailable. It is essential for good seed germination, photosynthesis, nitrate assimilation, proteosynthesis, fruit set and advancing maturity. It is not very mobile in the plant, so deficiency appears first on the youngest leaves.

Availability:

- Excessive levels of magnesium (> 18%) and/or humus (> 6%) block manganese absorption.
- A deficiency in sulfur (< 15 ppm) causes manganese absorption issues.
- A combined value of K + Na > 10% of CEC will reduce manganese absorption.

It is best to apply manganese to the soil while considering iron values (see section 20.2.10).

20.2.12 Molybdenum (Mo)

Molybdenum is highly important for nitrogen fixation in symbiotic relationships, particularly for legumes, and in crop rotations benefiting from free-living diazotrophs like Azotobacter. Below a threshold of 0.20 ppm, soil performance and autonomy diminish, making it more susceptible to nitrogen deficiencies since symbiotic nitrogen fixation by bacteria cannot complement fungal activity. Molybdenum is also crucial for protein synthesis in all plants. Brassicas are notably sensitive to molybdenum deficiency. Application should not exceed 120 grams of molybdenum per hectare per application.

20.2.13 Zinc (Zn)

Zinc is also involved in protein synthesis, photosynthesis, and sugar metabolism. It plays a crucial role in water use efficiency, akin to potassium. Regularly, we have observed soil fertilizations enriched with zinc significantly enhance crop and pasture resilience to summer drought and heat. It is one of the elements to monitor closely to maintain active vegetation during summer and reduce the climactic lag induced by our practices.

Availability: Heavy nitrogen fertilization and high levels of phosphorus and calcium significantly reduce zinc availability and uptake. Cold and humid weather conditions also negatively impact zinc absorption by plants.

Application: Soil applications are preferred to activate numerous soil microbial enzymes that depend on zinc and to supply the plant adequately. Zinc levels should always be higher than copper levels.

Soil application rates can reach up to 40 kg/ha/year of zinc sulfate.

20.3 Comfort zones and intervention zones

The comfort, vigilance and intervention thresholds for achieving soil mineral balance are described in Figure 20.4.

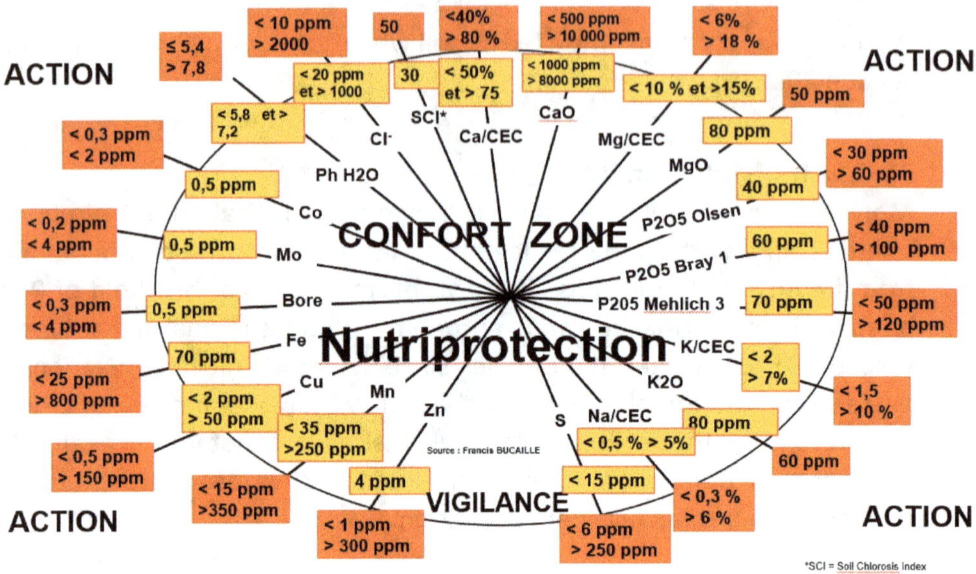

Figure 20.4 – Comfort, vigilance and intervention thresholds for achieving soil mineral balance

20.4 Managing organic fertilization

It is often recommended to apply exogenous organic matter based on soil analyses when nutrient levels are low. However, at this rate, 90% of soils would require amendment without assurance that nutrient levels will sustainably increase if storage mechanisms are not in place. In the book, you will find solutions to restore these soil storage functions autonomously, independently of external inputs such as manure. Autonomy in this matter is achievable.

Manure or compost amendments are excellent due to their role in both reseeding with beneficial microorganisms and the supply of complex mineral and organic molecules they contain. However, they also come with drawbacks. Their variable characteristics can sometimes exacerbate already unfavorable soil ratios. We have previously recommended temporarily halting organic matter applications due to excessive potassium levels in soils (over 20% of CEC). This is common on livestock farms where slurries and manures are spread on nearby fields. Therefore, they do not constitute the primary method for revitalization for us, although they can be a component in many situations.

Nevertheless, here are some good practices for managing manure and slurries that we recommend for soils:

- Recurrent applications should be validated through soil analyses to ensure that we are not exacerbating a pre-existing imbalance (especially excessive potassium), which could deteriorate soil structure, lead to accumulation of poorly decomposed organic matter, and foster unwanted flora such as bindweed, thistle, goosefoot, dandelion, and buttercup. In such cases, the "Law of the Minimum" applies.
- Leave any material that "smells" on the surface to allow complete oxidation of reduced molecules like ammonia, hydrogen sulfide, etc.
- Whenever possible, surface composting is a good practice (unless the goal is to destroy weed seeds). Surface composting, as opposed to windrow composting or anaerobic digestion, preserves all the energy contained in its organic components for soil fauna and microbiology, benefiting the entire food chain: carabides, insects, earthworms, actinomycetes, fungi, bacteria, viruses, etc.

Lastly, it's important to recognize that despite the undeniable qualities of composts, manures, and slurries, this category of products cannot rebalance a soil if it lacks calcium, magnesium, sodium, or a specific trace element, as their concentrations are too low to have a significant corrective impact.

20.5 Summary: managing chemical fertilization in five steps

Here we summarize, in order of priority, the five steps to mineral soil revitalization.

1. **Adjust Ca/Mg ratio.**
2. **Correct K+ and Na^+saturation.**

   > At the end of steps 1 and 2, the ideal pH of 6.3/6.8 will have been reached (variable according to soil type). The target H^+ saturation is 10%. Note : Slight acidity is beneficial for mineral bioavailability and fungal life.

3. **Correct sulfur and phosphorus levels.**
4. **Correct trace elements.**

   > By the end of step 4, optimal conditions for roots and soil life are in place, allowing nitrogen fertilization to become more effective.

5. **Consider nitrogen fertilization by choosing the right forms (see steering diagram).**

21 • SOIL REVITALIZATION IN 7 STEPS

In the hierarchical order of priorities, we summarize a soil revitalization itinerary in seven steps, incorporating various descriptions previously outlined in the book. The framework of bioindicator plants described in Chapter 16 of the book will help refine or confirm intervention decisions.

21.1 Restore soil verticality: identifying plow pans or no-till hardpans

Some compactions are so pronounced that even the most aggressive plants won't be able to develop a dense, deep root system and restore normal water functioning in a short space of time.

- **Method**: direct observation, using a spade and a binocular magnifying glass in the -35/-50 cm horizon (see description).
- **Solution**: crack if necessary. Once will suffice if other soil revitalization measures are implemented.

21.2 Restore the ratio of cations on the CEC

In line with the methodology described, this is the second most frequent and penalizing limiting point, since it's not just a question of pH. It's about soil physics: how much air and water can the soil hold? How will it resist compaction and erosion? Will it be able to support humifying fungal flora?

- **Method**: analyze soil, calculate cation occupancy on CEC and correct.
- **Solution**: apply fertilizers and soil improvers as indicated in § 19.2 and 20.1.

21.3 Provide other minerals necessary for soil enzymes

Minerals and trace elements serve first and foremost to activate the tools of soil microbiology: enzymes. Enzymes are the basic tools in the service of soil biodiversity and functionality.

- **Method**: soil analysis with measurements of all elements needed.
- **Solution**: add what's missing, taking into account the synergies and antagonisms of section 20.3.

21.4 Reintroduce families of microorganisms that may be lacking

Strains linked to the nitrogen cycle (diazotrophs, azotobacters) and mineral bioavailability (PGPR, phosphate solubilization and *Bacillus mucilaginosus*) are an essential cog in the nutrient process. Barring repeated soil sterilization (fumigation, steam, metam-sodium), there are still sufficient spores of mycorrhizae, basidiomycetes and ascomycetes in all soils.

- **Method**: microbiological laboratory analysis if necessary.
- **Solution**: supply *inoculums* according to manufacturer's instructions.

21.5 Cover the soil to best mimic the original climax

In temperate climates, the aim is mainly to photosynthesize in spring and summer. Return lignified vegetation to the soil in late autumn/winter. Avoid "lettuce", otherwise graze it or harvest it for fodder or silage. Avoid growing green manure to compensate for the sale and export of straw.

- **Method**: calculate and anticipate crop and cover crop maturity periods.
- **Solution**: change crop rotation if necessary. Adapt the earliness class of crops and cover crops. Modify the grazing and feeding plan.

21.6 Activate soil-humifying fungal flora

Many operations tend to inhibit this fragile flora. There are soil prebiotics capable of reviving this fraction of soil microbiology, which is the keystone of the carbon cycle.

- **Method**: observation of harvest residues. After a year, there should be nothing left.
- **Solution**: return woody material to the soil and apply a humifying prebiotic.

21.7 Reduce or eliminate all end-of-cycle fungicides – Practice nutriprotection

To mobilize the humification fungal flora around crop residues and build humus, it is vital to reduce or avoid the last fungicide, some of which is likely to end up in the straw, especially if maturity is accelerated by a hot spell. Maintain plant health with very light fungicide programs, or even totally supplanted by appropriate foliar Nutriprotection.

Conclusion

The great myths and major religions depict a deity creating humanity *ex nihilo* from clay. Recent scientific research from Cornell University establishes that clay, with its remarkable catalytic properties, may indeed have played a role in the synthesis of the core of life: the initial building blocks of primitive RNA (Cornell University, 2013). Science thus validates in a way the narrative of humanity crafted by the imagination and intuition of our ancestors. The convergence of spirituality and established scientific reality compels the Homo species to elevate its relationship with the soil. Whether Homo is sapiens (wise) or economicus (guided by rationality and optimal use of scarce resources), whether driven by wisdom or economic interest, Homo will benefit, ensure its survival, and perhaps leave future generations with the comfort of a restored Earth.

MAN, A LINK DEPENDENT ON ECOSYSTEMS

Within ecosystems, humans hold a unique place. As farmers, they disrupt the natural ecosystem and redirect it towards a nourishing function. When humans claim they "make plants grow," it's an inaccurate emphasis. Humans merely create conditions for a single plant species to dominate an area they themselves have delimited, known as a field. Humans do not make anything grow at all. It's the life force contained within the seed, the genes, enzymes, microbiology, and photosynthesis that express themselves. Humans disrupt, humans utilize nature. They are not above it; they are part of it, and their survival depends on it. This more modest view of humanity's place is crucial for adopting a more ecological approach to agriculture. A convergence of signals shows that humans are not the center of the world, but just a link. Without this awareness, it seems challenging to build a living agronomy. However, if we accept our position within an ecosystem, we can understand that every individual, including weeds and pests, is essential and provides services through their corrective activity and the signals they send us.

Cultivating plants adapted to human health?

> Aware of Gause's principle regarding niche breadth and dietary habits, we should also question how a cultivated plant that proves so well adapted to the nutrition of a pest could also constitute quality food for human health. Shouldn't we instead thank pests for preventing us from harvesting plants that are not suitable for our nutrition and well-being? For humans to be well-nourished and healthy, it is essential first and foremost that the plant itself is well-nourished. For the plant to be well-nourished, the soil must mobilize and transfer not only the thirty or so essential minerals but also its microbiology must provide plants with vitamins that some can only synthesize through rhizosphere bacteria. This is why we argue that humans thrive when the soil is alive. We are destined to cherish the life that soil harbors and to respect it.

During evolution, living organisms have tended to become more complex. However, this does not mean that simpler life forms have disappeared. Complex living beings also have complex ecological niche requirements. Some species of flowering plants considered among the most evolved, such as orchids, for example, depend on a single species of insect for pollination. Humans are among the relatively recent living beings on Earth who have found in ecosystems a diet rich in vitamins, minerals, sugars, and complex lipids. The degradation of the environment to a state preceding the development of soils simplifies soil food chains, favoring bacterial chains. These trophic chains lead to less nutrient-rich and less suitable food for human nutritional requirements. Maintaining a living soil is essential for humans to remain alive.

FARMERS HAVE A SPECIAL ROLE TO PLAY

The cultivated soil ecosystems represent hundreds of thousands of years of evolution. They belong more to the heritage and common good of all humanity than to individual familial legacies. Farmers have the power and privilege to manage this universal heritage that they steward. They also have the duty to pass it on to future generations. This tremendous power comes with a profound responsibility that few citizens bear: farmers have the ability to change the world, the model, and the relationship with living organisms on the 2, 20, 500, 1,000, or 10,000 hectares they cultivate.

RESTORING CONFIDENCE IN FARMERS

The new form of agriculture that lies ahead of us involves shifting from being mere operators to becoming managers of cultivated ecosystems. Direct and exclusive focus on yield and the production apparatus (the cultivated plant) must give way to a management approach centered around fertility in the broadest sense. This entails viewing soil as a full-fledged ecosystem. A significant part of the farmer's work will now involve creating conditions to restore the functionalities of this ecosystem. The cultivated system will always be somewhat disconnected from the natural system. The economic and ecological challenge is to reduce these discrepancies to maximize the benefits of the free functionalities associated with an efficient food chain in the soil.

To transform agriculture in this direction, we must embark on an "agricultural transition." Transition because it involves a technical path and human involvement. Support for change is necessary to progress step by step along a marked path that allows farmers to relearn how to move in a different direction. Throughout this journey, the focus should initially be on rebuilding solid foundations and shifting efforts away from parasitism, diseases, and pathogens. Instead, practices and research should primarily aim at managing these organisms through Nutriprotection and soil dietetics, rather than their annihilation. The focus should be on Life: soil life, ecosystem life, rhizosphere, and phyllosphere. The dual benefit is sustainable efficiency of agricultural systems and the construction of a more virtuous agriculture on solid foundations, without exposing farmers to financial risks. This process builds farmer confidence as they observe increasingly evident signs of crop health.

Quoting Sylvain Tesson from his "Eloge de l'énergie vagabonde" (In Praise of Wandering Energy): "I am wary of this term like all euphemisms; sustainable development is the balm applied to their guilty conscience by Westerners eager to continue enjoying without really lowering the world's fever. The term hides the wish to better adjust the reins to prolong humanity's race as long as possible."

MAN IS ALIVE WHEN THE SOIL IS ALIVE

For an agricultural ecology encompassing all hectares and all people

For us, ecology represents an awareness of humanity's place in the universe; it's not activism but a simple belief in Earth, in Gaia. Ecology begins with science and extends to generosity. It cannot be about organizing an artificial space, a living museum with 1,200 botanical species from around the world on half a hectare. It should not resemble a diorama of an imagined nature that never existed. Nor can it be reduced to a few hectares "biologically" fertilized by concentrating this "brown gold" (imported organic materials and straw collected from surrounding cereal fields like a survivalist in a bunker). It cannot be 30 goats grazing on a parched scrubland in summer, as this very practice has greatly expanded deserts globally. It cannot be a system constantly borrowing from creditors such as the forest (wood chips and compost), neighboring cereal farmers (straw), livestock farmers (manure, compost), or the earth's depths (peat). Globally, speaking in financial terms, this would consist of repaying annuities with new loans (overindebtedness). Under these conditions, it remains a practice rooted in mining and extraction, a pillage that lacks global virtue. The products used may be natural and healthy, but they do not form the basis of a sustainable system. It cannot be this way because none of these practices meets the global, humanitarian, magnanimous need to improve the state of the planet and the fate of humanity. The true measure is: can any of these models be applied sustainably to 100% of the world's agricultural land? If the answer is no, then everything is said.

Ecology — a science, not a philosophy or ideology — requires rising above, considering increased CO_2 levels, ozone concentrations, rising temperatures, sea level rise, global food supply, and human health. Ecology, by definition, is holistic: a few islands of nostalgia won't suffice to meet these global challenges. Local ventures that cannot be replicated globally cannot provide global solutions. **Every hectare of the world's cultivated land must be able to revitalize itself as its own source of fertility.**

INTENSIVE AND RENEWABLE USE OF RESOURCES IS PART OF ECOSYSTEMS.

Ecology, as we have seen in this work, cannot consist of simply being "against" something just because "we don't want that kind of agriculture"; it cannot consist of fighting against the creation of water reservoirs, against "large farmers," against "large herds," against intensification. Why not? Because Nature is our model, and it is intensive; because Nature is excessive. It is fashionable to worship the small, the "human scale." Of course, we need humanity, awareness, proximity, and communication. But that is not the point. Do you think the Amazon rainforest is reasonable, human-scale, or extensive? It is anything but that. The Amazon rainforest is a model of gigantism, excess, and intensification. Nowhere else do we have such productivity per hectare per year. Do you seriously think that the herds of wildebeests in the Serengeti, numbering in the millions, herds of zebras in the hundreds of thousands, hordes of antelopes of similar numbers, buffalo herds of thousands, and gatherings of bison in North America numbering in the millions (before their slaughter in the 19th century)—do you think these herds are human-scale or extensive? Certainly not. Anyone who has seen Kenya's buffaloes or North America's bison is amazed by their physical condition. Do you seriously think the crazy productivity of the Amazon forest on a continental scale and the enormity of African herds are endangering the planet? **The issue is not size but rather the mode of exploitation, compatibility with the original climax, and the revival of carbon, nitrogen, and water flows as they existed before deforestation.** Certainly, the small size of a farm can mask land and herd management errors by diluting their impacts. Certainly, it is more emotionally challenging to question a small producer. But what matters above all is identifying the predominant functions in these natural systems (biomimicry) that are atrophied or even annihilated when humans interfere. The farm size is only a revelation, not the cause. If size generates problems, it is because the mode of exploitation is not virtuous. Making it ecologically relevant and sustainable, that is the real battle. Farm size is a social, societal, political issue, certainly not an ecological one.

Climax determines maximum productivity

> The revitalization of soils and agroecosystems involves strengthening the natural functions that are failing in all agroecosystems worldwide, regardless of their size, by

- using climax as a model to restore maximum productivity.

All the ecosystems mentioned above share another characteristic besides their remarkable productivity: their state of health. Is the Amazon rainforest plagued by destructive waves of insects and bacterial fires? No ! Certainly, fungi, so-called pathogenic bacteria, viruses, and insects are present, but they are marginal and only fulfill their role as nature's recyclers within their strict ecological niche. Their presence is extremely beneficial. Are these ecosystems subjected to regular inputs of phytosanitary products? Not at all. They are healthy, autonomous, and resource-efficient. A living soil generates savings in fertilizers, crop protection products, and soil work. This is what we should strive towards.

SOIL REVITALIZATION IS A VIRTUOUS AND ECONOMICAL PROCESS

Revitalizing soils not only enhances the productivity and profitability of agrosystems but also promotes the health of plants, animals, and humans who consume them. Solutions to achieve this already exist, and several proven approaches are detailed in this book. These strategies are ready to be refined and expanded upon. Soil, being universally distributed on our planet, is powerful and rich in biodiversity, holding within it the capacity to provide answers to the global challenges humanity faces. We have the means to implement this virtuous cycle. Whether as a human community, a biological unit, or an economic actor, we will individually and collectively benefit from this approach. We have the capability to prove the doomsayers and collapse theorists wrong... provided that we make changes. So let's do it!

All evidence suggests that:

Man is and will remain alive if the soil is alive !

BIBLIOGRAPHY

Akin D. E., 1979, "Microscopic evaluation of forage digestion by microorganisms – a review", *J Anim Sci*, 48, 701-710.

Almeida Assunção S. et al., 2018, "Soil organic matter fractions affected by N-fertilizer in a green cane management in Brazilian Coastal Tableland", *Bragantia, 77* (2).

Arihara J. & Karasawa T., 2000, "Effect of previous crops on arbuscular mycorrhizal formation and growth of succeeding maize", *Soil Science and Plant Nutrition*, 46 (1).

Arnott H. J., Roseman T. S., Graham R. D. & Huber D. M., 1991, "An experimental study of Mn oxidation mineralization in the take-all fungus *Gaeumannomyces graminis* var. *trikini*", *Mycol. Soc. Am. News*, 42(3).

Arvalis – Institut du végétal, 2014, *Effets du travail du sol sur les cycles biogéochimiques (azote et carbone).*

Bailey V., Smith J. L. & Bolton H., 2002, "Fungal-to-bacterial ratios in soils investigated for enhanced C sequestration", *Soil Bio Biochem,* 34, pp. 997-1 007.

Baize B. et al., 2014, "Potentially Harmful Elements in Forest Soils" In Bini C. & Bech J. (eds), *PHEs, Environment and Human Health,* Springer, Dordrecht. https://doi.org/10.1007/978-94-017-8965-3_4.

Barak P. et al.1997, "Effects of long-term soil acidification due to nitrogen fertilizer inputs in Wisconsin" Plant and Soil, 197, pp. 61-69.

Bar-On Yinon M., Phillips R. & Milo R., 2018, "The biomass distribution on Earth.", *Proceedings of the National Academy of Sciences,* 115.25, pp. 6 506-6 511.

Bassem Dimassi et al., 2014. "Long-term effect of contrasted tillage and crop management on soil carbon dynamics during 41 years", *Agriculture, Ecosystems and Environment* http://dx.doi.org/10.1016/j.agee.2014.02.014.

Biao Zhu et al., "Rhizosphere priming effects on soil carbon and nitrogen mineralization", *Soil Biology and Biochemistry*, 76, pp. 183-192, 10.1016/j.soilbio.2014.04.033.

Callot G., 1999, *La truffe, la terre, la vie*, éditions QUAE, 210 p.

Carney K. M. et al., 2007, "Altered soil microbial community at elevated CO_2 leads to loss of soil carbon", *Proceedings of the National Academy of Sciences,* 104 (12), pp. 4990-4995, doi:10.1073/pnas.0610045104.

Chaboussou F., 1985, *Santé des plantes, une révolution agricole*, Flammarion [2011], "Les plantes malades des pesticides", Utovie.

Changfu H., Yiqi L. & Weixin C., 2017, "Rhizosphere priming effect: A meta-analysis", Soil, Biology and Biochemistry, 111, pp. 78-84.

Chikishev, A. G. (ed.) 1965. Plant indicators of soils, rocks, and subsurface waters. 210 pp. Consultants Bureau, NY.

Cohan J.-P. & Mary B., 2014, *Effet du travail du sol sur les cycles biogéochimiques de l'azote et du carbone*, Inra et Arvalis.

Commanday F. & Macy J. M., 1985, "Effect of substrate nitrogen on lignin degradation by Pleurotus ostreatus", *Arch. Microbiol.*, 142, 61-65.

Cornell University, 2013, "Clay may have been birthplace of life on Earth, new study suggests", *ScienceDaily*, disponible en ligne sur www.sciencedaily.com/releases/2013/11/131105132027.htm.

Crovetto C., 2000, *Les fondements d'une agriculture durable,* Teknea.

Daane L. L. et al., 1996, "Influence of Earthworm Activity on Gene Transfer from Pseudomonas fluorescens to Indigenous Soilé", *Bacteria Applied and environmental Microbiology*, 62 (2), pp.

515-521.

Davidson D. T., 1962, "Soil stabilization with chemicals", joint *Publication Bulletin*,193, et *Iowa Engineering Experiment Station Bulletin*, 22, Iowa Highway Research Board.

Detlef E. et al., 2005, "Environmental science – Carbon unlocked from soils", *Nature*, 437, 205-6. 10.1038/437205a.

Ducerf G., *Fascicule des conditions de levée de dormance des plantes bio-indicatrices*, Promonature, 2015.

Duchaufour Ph., 1950, "L'humus forestier et les facteurs de sa décomposition", *La revue Forestière*, 11.44.2, pp. 479-488.

Floudas D. et al., 2012 "The Paleozoic Origin of Enzymatic Lignin Decomposition Reconstructed from 31 Fungal Genomes", *Science*, 336 (6089), pp. 1 715-1 719.

Gause G. F., 1934, "The Influence of Biologically Conditioned Media on the Growth of a Mixed Population of Paramecium caudatum and P. aureliax", *Journal of Animal Ecology,* 3(2), pp. 222-230.

Glaeser S. P. et al., 2016, "Non-pathogenic Rhizobium radiobacter F4 deploys plant beneficial activity independent of its host Piriformospora indica", *ISME J.*, 10(4), pp. 871-884, doi:10.1038/ismej.2015.163.

Gleixner G. et al., 2001, "Plant compounds and their turnover and stabilization as soil organic matter", In Schulze E. D et al. (eds) *Global biogeochemical cycles in the climate system,* pp. 201-215.

Gleixner G., 2013, "Soil organic matter dynamics: A biological perspective derived from the use of compound-specific isotopes studies", *Ecological Research,* 28, 10.1007/s11284-012-1022-9.

Grenet E., 1970, "Taille et structure des particules végétales au niveau du feuillet et des fèces chez les bovins", *Ann. Biol. Anim. Bioch. Biophys.,* 10, pp. 643-65.

Griffin D. M & Quail G., 1968, "Movement of bacteria in moist particulate systems", *Aust J Biol Sci*, 21 pp. 579-582.

Grummer G., 1961, "The role of toxic substances in the inter-relationships between higher plants", *Symp. Soc. Exp. Bio.,* 15, pp. 219-228.

Gui Li Li, 2019, "Interactions between microorganisms and clay minerals: New insights and broader applications", *Applied Clay Science*, 177(1), pp. 91-11.

Heckman J. R., Clarke B. B. & Murphy J. A., 2003, "Optimizing manganese fertilization for the suppression of take-all patch disease on creeping peinturasse"., *Crop Sci.*, 43, pp. 1395-1398.

Hodge, A., Campbell, C. D., and Fitter, A. H., 2001. "An arbuscular mycorrhizal fungus accelerates decomposition and acquires nitrogen directly from organic material". *Nature* 413, 297-299.

Hoorman J. J. et al., 2011, *The biology of soil compaction Leading Edge*, pp. 584-587.

IAEA, 2006, *Environmental Consequences of the Chernobyl Accident and Their Remediation : Twenty Years of Experience – rapport of the UN Chernobyl Forum Expert Group "Environment"*, Radiological Assessment Report Series.

Isichei A. O., 1990, "Arid Soil Research and Rehabilitation", *The role of algae and cyanobacteria in arid lands. A review*, 4 (1).

Isichei A. O., 1990, "The role of algae and cyanobacteria in arid lands. A review", *Arid Soil Research and Rehabilitation*, 4 (1).

Jameel, Q. Y., & Mohammed, N. K. 2021" Protective rules of natural antioxidants against gamma-induced damage—A review. *Food Science & Nutrition*, 9, 5263– 5278

Jarrige R., 1966, *Nutrition des ruminants domestiques. Ingestion et digestion INRA,* Editions 1995, pp. 607.

Jarrige R., 1980, *Nutrition des ruminants domestiques*, INRA Editions 1995, pp 631.

Johnson D., 2015, PeerJ PrePrints, http: //dx.doi.org/10.7287/peerj.preprints.789v1.

Joung Y. S., 2015, "Aerosol generation by raindrop impact on soil", *Nat Commun,* 6, p. 6 083, https://doi.org/10.1038/ncomms7083.

Kashif B., 2012, *Effect of different phosphorus levels on growth and nitrogen fixation.*

Kellogg C. A. & Griffin D. W., 2006, "Aerobiology and the global transport of desert dust", *Trends Ecol Evol* 21, pp. 638-644, doi :21. 638-44. 10.1016/j.tree.2006.07.004.

Khalil M. I. et al., 2007, "Nitrogen fertilizer–induced mineralization of soil organic C and N in six contrasting soils of Bangladesh", *J. Plant Nutr. Soil Sci*, 170, pp. 210-218.

Klironomos J. N., 2003, "Variation in plant response to native and exotic arbuscular mycorrhizal fungi", *Ecology*, 84, pp. 2 292-2 301.

Kohlmeier S. et al., 2005, "*Taking the fungal highway: Mobilization of pollutant-degrading bacteria by fungi*", *Environmental science & technology*, 39.

Krupnik T. J., Six J., Lada J. K., Paine M. J. & van Kessel C., 2004, "An assessment of Fertilizer Nitrogen recovery Efficiency by grain crops", *Agriculture and the nitrogen cycle*, pp.195-196.

Labreuche et al., 2001 [Educagri éditions 2016], cité dans " Gestion de l'état organique des sols avec SIMEOS AMG ", Tomis V., Agro-Transfert RT, Conférence état organique des sols 9 février 2015.

Langley J. A. et al., 2009. "Priming depletes soil carbon and releases nitrogen in a scrub-oak ecosystem exposed to elevated CO_2", *Soil Biology & Biochemistry*, 41, pp. 54-60.

Lastuvka Z., 1955, "Effect of couch grass on growth of wheat and rye", Cˇs. *Biol*, 4, pp. 165-175.

Leatham G. F. & Kent Kirk T., 1983, "Regulation of ligninolytic activity by nutrient nitrogen in white-rot basidiomycetes", *FEMS Microbiology Letters,* 16(1), pp. 65-67.

Leatham G. F & Kirk T. K., 1983, "Regulation of ligninolytic activity by nutrient nitrogen in white-rot basidiomycetes", *FEMS Microbiology Letters*, 16 (1).

Lecomte P. et al., "Maladies du bois de la vigne", dans Inrae, disponible en ligne sur www.maladie-du-bois-vigne.fr/Diffusion-des-resultats/Lecomte-P-et-al.-2019.

Leininger S. et al., 2006, "Archaea Predominate Among Ammonia-Oxidizing Prokaryotes in Soils", *Nature*, 442, pp. 806-809.

Levi M. P. & Cowling Met E. B., 1969, "Rôle of nitrogen in wood deterioration", *Phytopathology*, 59, pp. 460-468.

Malik et al., 2016, "Soil Fungal:Bacterial Ratios Are Linked to Altered Carbon Cycling", *Frontiers in Microbiology*, 7, art. 1247.

Mangenot J. F., 1980, "Les litières forestières : signification écologique et pédologique", *Revue forestière française*, 4.

Mathre D. E., 1995, "Gaeumannomyces", in *Methods for Research on Soilborne Phytopathogenic Fungi.*, APS press.

Pausch J. et al., 2013, "Weixin Cheng Plant inter-species effects on rhizosphere priming of soil organic matter decomposition", *Soil Biology and Biochemistry*, 57, pp. 91-99.

Phillips R. P. et al., 2012, "Roots and fungi accelerate carbon and nitrogen cycling in forests exposed to elevated CO_2", *Ecology Letters*, 15, pp. 1 042-1 049.

Read D. J., 1991, "Mycorrhizas in ecosystems – nature's response to the "law of the minimum"", *Frontiers in mycology*, p. 101-130.

Reid I. D., 1979, "The influence of nutrient balance on lignin degradation by the white-rot fungus Phanerochaete chrysosporium", *Can J Bot*, 57, 2 050-2 058.

Rillig M. C. et al., 1999, "Rise in carbon dioxide changes soil structure", *Nature*, 400, p. 628.

Rusch H. P. 1968 « *Bodenfruchtbarkeit* » *(« Fertility of the ground »)*, published by K. F. Haug Verlag.

Sampson, A. W. 1939. Plant indicators-- concept and status. Bot. Rev. 5: 155-206

Sartre J.-P., 1943, *L'Être et le Néant*, Gallimard.

Saverio Fiore, 2011, "Bacteria-induced crystallization of kaolinite", *Applied Clay Science*, 53(4), pp. 566-571.

Schimel J. P. & Schaeffer S. M., 2012 , "Microbial control over carbon cycling in soil", *Front Microbiol.*, 3, p. 348, doi:10.3389/fmicb.2012.00348.

Schulze E. D. & Freibauer A., 2005, Environmental science: Carbon unlocked from soils", *Nature*, 437 : 205-206.

Shahzad T. et al., 2012, "Role of plant rhizosphere across multiple species, grassland management and temperature on microbial communities and long term soil organic matter dynamics",. Agriculturalsciences. AgroParisTech.

Singleton L. L., Mihail J. D. & Rush C. M. (eds), 1992, *The American Phytopathological Society*, St. Paul, MN, pp.60-63.

Sinsabaugh R. L. et al., 2013, "Carbon use efficiency of microbial communities: stoichiometry, methodology and modelling" *Ecology letters*, 16(7), pp. 930-939.

Six J. et al., 2006, "Bacterial and fungal contribution to carbon sequestration in agroecosystems", *Soil Sci. Soc. Am. J.*, 70, pp. 555-569.

Song Y., 2015, "Hijacking common mycorrhizal networks for herbivore-induced defence signal transfer between tomato plants", *Scientific Reports*, 4, article 3915.

Strickland M. & Rousk J., 2010, "Considering fungal: Bacterial dominance in soils - Methods, controls, and ecosystem implications", *Soil Biology and Biochemistry*, 42, pp. 1 385-1 395. 10.1016/j.soilbio.2010.05.007.

Timonin M. I., 1972, "Oxidation of manganous salts of manganese by soil fondit", *Revue canadienne de microbiologie*, 18 (6), pp. 793-799.

Vries F.T., 2009, *"Soil fungi and nitrogen cycling. Causes and consequences of changing fungal biomass in grasslands"*. Ph.D thesis Wageningen Universiteit, Wageningen, ISBN 978-90-8585-325-1 https://edepot.wur.nl/1757.

Waring B. G. et al., 2013, "Differences in fungal and bacterial physiology alter soil carbon and nitrogen cycling: insights from meta-analysis and theoretical models", Ecol. Lett., 16, pp. 887-894 10.1111/ele.12125.

Warmink J. et van Elsas J., 2009, "Migratory Response of Soil Bacteria to Lyophyllum sp. Strain **Karsten** in Soil Microcosms", Applied and environmental microbiology, 75, pp. 2 820-2 830. 10.1128/AEM.02110-08.

Yang P., 2017, *Mechanisms of migration of Paraburkholderia terrae BS001 in the mycosphere*, University of Groningen, disponible en ligne sur http://hdl.handle.net/11370/9d788301-001d-4de2-8775-41010de0dc30.

www.ingramcontent.com/pod-product-compliance
Lightning Source LLC
LaVergne TN
LVHW052309200926
848481LV00002B/506
* 9 7 9 1 0 9 7 9 4 1 3 0 7 *